Copyright Year: 2008.
Copyright Notice: by Qaqish & Kiatoukaysy.
All rights reserved.

ISBN: 978-0-6151-9759-3

Copyright notice: © by Qaqish & Kiatoukaysy

www.StepBooster.com

Disclaimer: The information presented in this book has been carefully reviewed and to the best of the authors knowledge, is correct. Please note that any information found to be incorrect was done so inadvertently; neither the publisher nor authors will accept any legal responsibility for material presented in this book. The information presented in this book is to aid in passing USMLE Step 2 and 3 examinations only and is not to be used as a source for the actual treatment of patients in real life settings.

The USMLE Step Booster

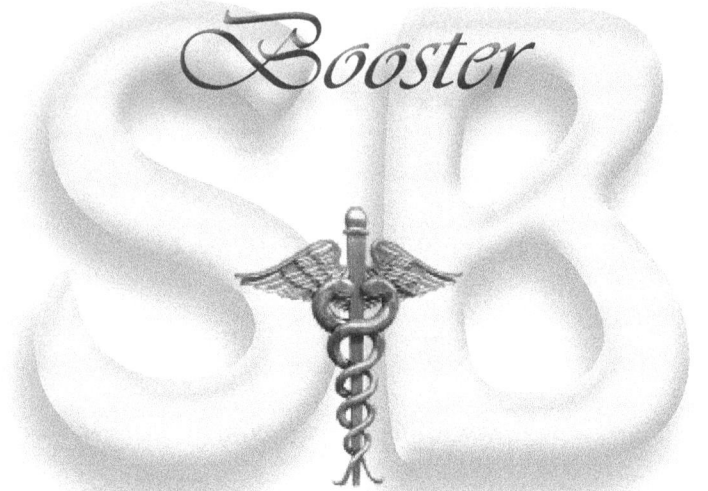

To all my friends and family,
to my parents, Seif and Laila Qaqish, and to my brother Laith...
thank you for believing in me throughout the years!

Saed Qaqish, MD

To my parents, friends and family... without you I would not have been
able to succeed. Thank you!

Linda Kiatoukaysy, MD

The USMLE *Step Booster*

PREFACE

Most publications reviewing the USMLE claim that their books are complete with all the material needed for you to do well on the exam. I do not underestimate their strengths nor deny that many of the famous titles out there are extremely helpful and sworn by! Nonetheless, examinees are always searching and hunting for the perfect book, the one that has it all. Unfortunately, due to the vast array of medical knowledge required to pass the United States Medical Licensing Examination[R] most often more than one resource is needed in order to pass and do well.

This book is not meant to be the one and only source that covers all the STEP 2 and 3 material... it's not your savior, not the answer to your prayers! But I promise you this: If you use this book as directed along with your study material of choice, you will most definitely be in better shape for the exam; you will be more prepared than ever and the proof will show on your score report.

Study hard, try to relax from time to time...
This too shall pass!

Thank you for buying this book.

Saed Qaqish, MD

The USMLE *Step Booster*

HOW TO USE THIS BOOK

The book is divided into 20 sections. Each section has randomized high yield questions and topics designed to be reviewed and memorized in one day. Read the questions, think about them and then look at the corresponding page for explanations. By the end of the day, you should be able to answer those questions without peeking at the answers.

The book is designed in a way that not only optimizes your study time but will aid in your ability to memorize important information. We recommend you study each section every day for an hour or so in the morning and then in the evening. The next day, do the same for the following section and so on. After studying three sections (3 days) you'll see a page noting (1 day break) giving you time to go back and review all the questions you've memorized thus far.

It seems like a light load, but after finishing all the sections, you will be surprised at the tremendous amount of information you have engraved into your brain from this simplistic yet efficient review!

Keep this book with you all day and night and whenever you get a chance to peek in, please do so!

Linda Kiatoukaysy, MD

SB

Day 1 — Questions 1/5

1. Difference between formula milk and regular milk?
2. Effect of Calcium and antacids on Iron absorption?
3. Effect of acids and vitamin C on Iron absorption?
4. Bleeding after dental procedures; bleeding time is elevated but everything else including platelet count is normal. BUN/Cr ratio is also normal. What is the cause?
5. What if the BUN/Cr is abnormal?
6. What test do you get if you're suspecting VW disease?
7. What is the genetic trait in VW disease?
8. Treatment of VW disease?
9. What is the relationship between phenytoin and oral contraceptives?
10. Drug of choice in the empirical treatment of a cystic fibrosis patient with severe exacerbation of pulmonary disease?

Day 1   Answers 1/5

1. Formula has more Iron! So if a child comes to your office after being switched to regular milk and you find out that he is anemic, advise the parents to increase the child's solid food intake.

2. They decrease Iron absorption. Tell your patients to take their Iron pills a couple of hours before or 4-5 hours after a meal.

3. Acids increase the absorption of Iron.

4. Von Willebrand's Disease (VW).

5. If the BUN/Cr is abnormal, then the bleeding is caused by Uremia affecting the platelet function.

6. Measure Factor VIII antigen and risocetin cofactor activity. If it's VW both will be low.

7. Autosomal Dominant.

8. Depends on type of bleeds:
- Mild bleeds: treat with Desmopressin (given only once in 24 hrs).
- Severe bleeds (prior to surgery or if Desmo was given but was Ineffective): treat with factor VIII concentrate.

9. Oral contraceptives do not have any affects on phenytoin but phenytoin enhances the metabolism of the oral contraceptives which can cause failure in contraception.

10. An Aminoglycoside with an anti-pseudomonal Penicillin (like Pipercillin and Tobramycin).

Day 1 — Questions 2/5

Topic: Human Bites.

1. What organism is usually the cause of the soft tissue infection in a human bite site?
2. How do you treat the infection?
3. What about animal bites? How do you treat those?

◊

4. How do you treat a fungal infection involving hair or nails?
5. Side effect of Ketoconazole?
6. What is the best empirical treatment for community acquired pneumonia?
7. When do you need to give steroids in an HIV patient with PCP?
8. How do you diagnose Chlamydia?
9. How do you treat Chlamydia?
10. A patient with Hepatitis complains of a movement disorder (chorea) and psychosis (confused). What diagnosis should you think of?
11. How do you confirm the diagnosis?
12. Treatment?

Day 1   Answers 2/5

1. Eikenella Corrodens (a gram negative anaerobe).
2. Ampicillin – Sulbactam.
3. Pasteurella Multocida is usually found in that case but the treatment is the same.
4. Terbinafine or Itraconazole.
5. Gynecomastia, and in the elderly it causes confusion.
6. Macrolides: Azithromycin, clarithromycin, or erythromycin. 2nd line: levofloxacin, gatifloxacin or moxifloxacin.
7. When it is a severe infection: $PO_2 < 70$ and A-a gradient > 35.
8. By getting a serology or fluorescent antibody testing.
9. Doxycycline.
10. Wilson's disease.
11. labs: low ceruloplasmin level, high urinary copper and elevated hepatic copper.
12. Penicillamine.

Day 1   Questions 3/5

1. If serum protein electrophoresis shows an IgM spike but no renal failure and bone x-ray is normal. What is the diagnosis?
2. What symptoms would they have?
3. What's the treatment?
4. What do you know about Heparin Induced Thrombocytopenia Type I:
5. What do you know about Heparin Induced Thrombocytopenia Type II:
6. How do you diagnose Type II?
7. How do you treat Type II?

        Topic: side effects of chemotherapy:

8. Adriamycin, doxyrubicin, daunarubicin?
9. Busulfan?
10. Cyclophosphamide, Ifosphomide?
11. Methotrexate?
12. Vincristine?

◊

13. What the normal hematocrite level in a newborn?
14. Why is it that you would get a Monospot test in children that are over 5 years old as compared to EBV titers in those younger than 5 years when you want to confirm mononucleosis?

Day 1   Answers 3/5

1. Waldernsrom's Macroglobinemia.
2. hyper viscosity symptoms: dizziness, visual disturbance, blurry vision.
3. emergency plasmaphoresis.
4. Non-Immune: occurs soon after starting therapy, usually mild and self limited.
5. Occurs 5-15 days after starting heparin. Patients are at an increased risk of thrombosis due to heparin induced platelet aggregation.
6. Order 14C serotonin release assay. Anti heparin/platelet factor 4 antibody.
7. Stop all forms of Heparin. If anticoagulation is still needed, use Danaproid, Lepirudin, Hirudin or Argatroban.
8. Cardiotoxicity.
9. Pulmonary fibrosis.
10. Hemorrhagic cystitis (use Mesna to prevent).
11. Liver toxicity.
12. Peripheral neuropathy.
13. 56%
14. Because the monospot test gives many false negative results when used in children less than 5 years old.

Day 1 Questions 4/5

Topic: when can I take my child back to school/daycare?

1. Measles?
2. Rubella?
3. Scarlet fever?
4. Impetigo?
5. Herpes Simplex infection?
6. Coxsackie virus infection (the hand foot and mouth disease)?
7. Varicella-Zoster (chicken pox)?
8. Hepatitis A?

◊

9. How does Roseola present?
10. What is the most common complication do children get due to Measles?
11. History of fever lasting more than five days with no known cause in a child with swollen palms and feet, a strawberry tongue and unilateral lymphadenopathy. What diagnosis should you think of?
12. How do you know it's not Scarlet Fever since they both present with strawberry tongues?
13. Treatment of Kawasaki?

Day 1   Answers 4/5

1. 4 days after the start of the measles rash.
2. 7 days after the start of the rash.
3. >24 hours after the initiation of therapy or after the fever is gone.
4. just like Scarlet fever, 24 hours after the initiation of the proper antibiotic treatment.
5. just cover up the rash and send back to school. You need to tell your child not to scratch it while in school.
6. it looks pretty bad, but the child should be in school! No need to take time off; it's not necessary unless very severe and the child cannot tolerate oral fluids.
7. The child should not go back to school until all the lesions have crusted and dried up; anytime before that, the child would be very contagious.
8. 7 days after the onset of the Hep A symptoms.
9. Patients are usually 6 to 15 months old. High grade fever for 3 days and a rash starts as soon as the fever is gone. Rash starts on the trunk and spreads out.
10. Otitis Media!
11. Kawasaki (mucocutaneous lymph node syndrome).
12. Scarlet has bilateral cervical lymphadenopathy (Unilateral in Kawasaki).
13. You should always get an echo to make sure they do not have a coronary aneurism! Then treat with IV Ig plus a high dose of Aspirin.

Day 1 Questions 5/5

Topic: Paroxysmal Nocturnal Hemoglobinuria (An acquired cell disorder).

1. What labs do you need in order to diagnose?
2. How do patients present?
3. Screening tests?
4. Confirmation test?
5. Treatment?

Topic: Treatment of lead Poisoning.

6. Acute:
7. Mild or moderate (lead level 50 – 70):
8. Severe toxicity (lead level > 70):

Topic: Side Effects of Herbs!

9. Aconite:
10. Kava:
11. Ginko Balboa:
12. Ginseng:

◊

13. What do you do when you have a pregnant patient with an elevated AFP? (Normal AFP is <2.5 MoM)

Day 1   Answers 5/5

1. Pancytopenia + hemolytic anemia.
   - Urine Hemosiderine will be present.
   - Low Alk. Phos.
2. May present with abdominal pain secondary to mesenteric or hepatic vein thrombosis.
3. Sucrose lysis test or Ham test.
4. Flow cytometry (you'll see a missing CD 59)
5. Prednisone + androgen (Donazol or Danocrine).
6. Activated charcoal plus cathartic.
7. Use an oral chelator (succimer)
8. EDTA plus BAL (Dimercaprol)
9. Cardiotoxic.
10. Causes liver injury.
11. May cause bleeding.
12. Psychosis and Steven-Johnson.
13. If there's a family history of neural tube defects, get an ultrasound.
- If AFP is >7.0 MoM, get an ultrasound.
- If in the above 2 scenarios the ultrasound came back normal, you need to get an Amniocentesis for further evaluation.
- If AFP is >2.5 but <7.0 MoM and there's no risk or history of neural tube defects, just repeat the maternal serum AFP.

Day 2 — Questions 1/5

Topic: Congenital Adrenal Hyperplasia:

1. What is the most common cause?
2. How do you confirm the diagnosis?
3. What's the difference between 21-hydroxylase deficiency and 11-hydroxylase deficiency?

◊

4. Your post-op patient is bleeding easily, PT and PTT are elevated. What should you suspect? How do you treat it?
5. If you suspect a Pulmonary embolism in a patient but the V/Q scan or spiral CT said "equivocal" what do you do next?
6. How do you treat if positive?

Topic: Otitis media.

7. If a child comes in with otitis media, how do you treat it?
8. What if the child had otitis media and conjunctivitis at the same time?

◊

Day 2　　　　　　Answers 1/5

1. A deficiency in 21-hydroxylase.

2. confirmation is the elevated levels of 17 alpha-Hydroxyprogesterone.

3. 11-hydroxylase deficiency is associated with hyponatremia.

4. Vitamin K deficiency (remember, this patient is post-op which means he has been NPO and is on broad spectrum antibiotics). Treat simply by giving vitamin K.

5. Get a Doppler ultrasound for both legs, if that's positive, then the diagnosis is confirmed. If negative and you still think it's a PE, do a pulmonary angiogram (which is actually the gold standard to diagnose a PE)

6. IV heparin followed by warfarin for 6 months. Keep INR between 2 and 3. Continue with low molecular weight heparin as an outpatient.

7. Amoxicillin should be able to treat most cases but if you give amoxicillin and the child doesn't improve after 24 hours, give Erythromycin and a sulfa instead.

8. in that case, Amoxicillin alone would not be good enough… you need to give Amoxicillin plus clauvulinic acid (the causative organism is most likely to be H.influenza in this case).

Day 2 — Questions 2/5

Topic: Hemochromatosis:
1. What test is used test to screen for hemochromatosis?
2. Test used to confirm the diagnosis?
3. Treatment of hemochromatosis?

Topic: lots of RBCs
4. How do you differentiate between polycythemia vera and relative polycythemia?
5. What's the treatment of polycythemia vera?
6. What's the difference between polycythemia vera and secondary polycythemia?
7. Common physical exam finding in polycythemia vera?

Topic: Cannon A waves.
8. What do you know about cannon A waves?
9. what is the best initial treatment for this problem?

◊

10. What do you do if a PAP smear came back saying ASCUS (atypical squamous cells of unknown significance)?

Day 2   Answers 2/5

1. Transferren saturation > 50% and elevated ferritin level.
2. Gene study.
3. Phlebotomy: start treating if ferritin is > 200 in pre-menopausal women or > 300 in men or post-menopausal. Deferoxamine is used if the patient cannot tolerate phlebotomy.
4. In polycythemia vera, you would see an increased red cell mass! Relative polycythemia has a normal RBC mass.
5. Phlebotomy to maintain Hct <45%
6. Erythropoietin is normal in vera.
7. Splenomegally.
8. Pulse waves in the neck.
- It means that the atria are pumping against a closed tricuspid valve.
- Always assume 3rd degree (complete) heart block even before getting an EKG.
- On EKG you would see P-waves with no association to QRS.
9. First give Atropin.
- If Atropin doesn't work, put a transcutaneous pacer.
- If all fails, use Dopamine.
10. Get an HPV DNA test.
- If HPV DNA was positive for high risk HPV, you should do colposcopy immediately.
- If HPV DNA is negative, you can just repeat the PAP in one year.

Day 2 — Questions 3/5

Topic: Hodgkin's lymphoma.

1. How do you stage Hodgkin's?
2. How do you treat?

Topic: Treatment of Gout.

3. How do you treat if the problem is over production of uric acid (urine uric acid >800)?
4. What if the problem is due to under excretion of uric acid (urine uric acid <800)?
5. How do you treat acute gout?

◊

6. list all post surgical complications possible after TURP (transurethral resection of the prostate).

Topic: Addison's disease.

7. Screening test?
8. Confirmation test?
9. Findings in Addison's?

Day 2 — Answers 3/5

1. Stage I: one lymph node involved.
- Stage II: two lymph nodes or more but on the same side of the diaphragm.
- Stage III: lymph nodes involved on both sides of the diaphragm.
- Stage IV: when it becomes a disseminated disease.

A: No symptoms other than a painless lymphadenopathy.

B: Symptoms include fever and night sweats.

2. IA, IIA get treated with radiation therapy.
- All other stages get treated with chemotherapy (ABVD: adriamycin, bleomycin, vincristin and dacarbazine)

3. Allopurinol.
4. probenicid.
5. Indomethacin or colchicine or if all fails, you can use steroids.
6. Hyponatremia: just an acute problem.
- Hematuria: usually gone by 3 weeks.
- Elevated levels of PSA: should return to normal within 4 weeks.
- Retrograde ejaculation: long term and occurs in 70% of post-TURP patients.
7. Cosintropyn stimulation test.
8. Morning serum cortisol <3.
9. low cortisol, aldosterone, Sodium and blood pressure. Increased Potassium levels and increased skin pigmentations.

Day 2   Questions 4/5

1. If you are giving Ferrous Sulfate to treat a patient with Iron deficiency anemia but the patient cannot tolerate the treatment due to side effects, what can you do?
2. What test would you want to obtain in order to monitor the progression of the patient's response to iron replacement in 2 weeks after starting the therapy?
3. Causes of Folate deficiency?
4. What labs do you see in Hemolytic anemia?

Topic: treatment of bradycardia.

5. What if symptoms are present?
6. What if no symptoms are present?

◊

Topic: chicken pox
(remember, back to school after all lesions have and dried up)

7. If a patient is older than 13 years but never had a varicella vaccination, what do you do?
8. What's the incubation period for varicella-zoster?
9. If an immunocompromised patient or a pregnant woman were exposed to varicella, what do you need to do?
10. What about Acyclovir, when do you give that?

Day 2 — Answers 4/5

1. You can replace Ferrous sulfate with Ferrous Gluconate which has lower elemental iron but is better tolerated by patients.
2. Reticulocyte count.
3. Methotrexate, Bactrim, sulfasalazine, Phenytoin and alcohol.
4. Increased reticulocyte count, increased LDH, increase indirect bilirubin and low haptoglobin.
5. Use Atropine then put on pacer.
6. Get an EKG and treat according to results:
- If sinus rhythm: no treatment needed.
- If first degree heart block: no treatment needed.
- If if second degree heart block:
- Mobitz type 1: no treatment needed.
- Mobitz type 2: treatment needed (even if no symptoms are present!!)
- If third degree heart block: treat even if no symptoms are present.
7. You give 2 doses of the vaccine about 4 weeks apart.
8. 2-3 weeks.
9. Give VZIg (varicella-zoster Immunoglobulin.
10. Only given to high risk patients that developed chicken pox even though you gave them VZIg.

Day 2   Questions 5/5

Topic: pituitary tumors.

1. How do you treat pituitary tumors?
2. What if the prolactinoma patient cannot tolerate Bromocriptine?
3. What if your patient is on Bromocriptin (or cabergolin) and became pregnant?
4. What do you know about the thyroid storm?
5. How do you treat a thyroid storm?

◊

6. What do you expect the Acid Base to be like in a patient who overdosed on salicylates?
7. Findings in Multiple Myeloma?
8. Treatemnt of Multiple Myeloma?

Day 2   Answers 5/5

1. if the patient is symptomatic, you treat them with surgery unless the problem is a prolactinoma (prolactin level >200) which is easily treated with Bromocriptin (a dopamine agonist).
2. You can use Cabergolin instead, and if the patient cannot tolerate cabergolin either, send them to surgery.
3. Not good! STOP the medication and just monitor the patient for enlargement of the pituitary by performing routine visual field exam.
4. Patients have symptoms of thyrotoxicosis, Atrial fibrillation, fever, tachycardic, diarrhea (every organ is hyper!) patients are delirium or have some type of change in mental status.
5. PTU + propanolol + Iodine + hydrocortisone.
6. Metabolic acidosis with respiratory Alkalosis.
7. Monoclonal "M" spike on protein electrophoreses.
- Lytic lesions shown on bone x-ray.
- At least 25% plasma cells on bone marrow biopsy.
- Renal insufficiency (due to the precipitation of Bence-Jones proteins produced by the plasma cells).
- Anemia (due to plasma cells/blast cells invading the bone marrow).
8. Prednisone, Melphenon then maybe bone marrow transplant. You can treat their hypercalcemia with IV fluids, Amidronate, diuretics and calcitonin.

Day 3 Questions 1/5

1. What do you expect the BUN/Cr ratio to be in a patient who has intestinal bleeding?
2. Treatment of Aspirin overdose?

    Topic: Hepatitis B and having a baby!

3. When do you give Hbig (hepatitis B immunoglobulin) along with the Hep B vaccination to a newborn?
4. How do you give it?
5. What if the mother's HBAg status is unknown at the time of delivery?

    Topic: dancers sign.

6. If on abdominal examination of a child patient you feel a sausage-shaped mass or if on the exam they mention "Dancers sign" what should you think of?
7. In what age group does the above presentation appear most commonly?

Day 3   Answers 1/5

1. BUN is elevated because of the absorption of Nitrogenous breakdown of RBCs from the GI tract.
2. Activated charcoal so the body doesn't absorb more of the drug.
   - IV fluids.
   - IV HCo3 to alkalinize the urine (to enhance the drug's excretion)
   - Get a psychiatric evaluation because it's not easy to overdose on aspirin by mistake.
3. ONLY if the mother is HBsAg negative!
4. it needs to be given within the first 12 hours to be most effective.
5. First, give the Hep B vaccine to the baby, then draw blood from the mother to check the Hep B status. If it comes back positive give Hblg (but no later than 7 days after the delivery).
6. Dancers sign means empty right lower quadrant. You should be thinking of Intussusception. Remember, those patients would have currant jelly stool (not cherry red as in Meckel's diverticulum) and Intussusception is painful while meckel's is painless.
7. Between 3 months and 3 years of age; if younger or older, think of something else.

Day 3   Questions 2/5

Topic: fever, headache and a stiff neck!

1. We all know it could be meningitis, but what do you do first in management?
2. When do you start the treatment?
3. What if you do need to get a CT? would you give treatment before or after CT?
4. What antibiotics do you use?

Topic: hyperthyroidism and pregnancy.

5. A pregnant patient comes to your office thinking that she has hyperthyroidism, what tests would you order for her and why?
6. What if her TSH was lower than 0.01 mµ/L and her T4 was elevated?
7. How do you diagnose GGT (gestational transient thyrotoxicosis)?

◊

8. How do you treat PMS/PMDD (premenstrual dysphoric disorder)?

Day 3   Answers 2/5

1. Get a lumbar puncture UNLESS there are signs of increased ICP or if there are focal neurological deficits present or if the patient is confused or unconscious··· in that case, you need to get a CT scan.
2. if there is no need for CT, get the lumbar puncture done, send for cultures and start treatment as you wait for culture results.
3. Give treatment before sending for CT.
4. Start with IV ceftriaxone and when the culture comes back, if the causative organism is sensitive to Penicillin, treat with Ampicillin. If not sensitive to penicillin, then keep on giving the ceftriaxone or give vancomycin.
5. Get free T4, total T4 and TSH levels. In pregnancy, serum concentration of TBG increases which would increase the levels of T3 and T4. In order to know if a pregnant patient really has hyperthyroidism you need the levels of free T4 and TSH.
6. Hyperthyroidism.
7. The patient would have a mild decrease of TSH at the same time of having an increased T4 and T3 levels.
- The problem is due to inceased levels of hCG at 8 to 14 weeks of gestation.
- About 10% of all pregnant women can develop this problem.
8. SSRIs, especially Fluoxetine. If your specific patient is refractory to Fluoxetine, then you would need to try Alprazolam instead.

Day 3   Questions 3/5

1. What is SMR and how do you calculate it?

    Topic: Bernard – Soulier Syndrome:

2. What is it?
3. How do the platelets look on the smear?
4. So what exactly is the problem?
5. How do you treat this?
6. What else should you know about Bernard Soulier?

    Topic: Jugular Venous Distention with Tachycardia and low Blood Pressure:

7. What is the diagnosis if crackles were heard on lung auscultation?
8. What if the lung sounds were clear but the patient had pulses paradoxes? How would you treat that?

    ◊

9. What drug would you use if your psychiatric patient is very combative and agitated and is about to harm self or others?
10. If you had a patient who is at high risk of developing a DVT, what should you do?
11. What should you do if a perimenopausal patient comes to you complaining of having irregular, heavy or dysfunctional vaginal bleeding?

Day 3 — Answers 3/5

1. SMR: standardized mortality ratio.
   - Calculated by dividing the observed number of deaths by expected number of deaths.
   - It represents an adjusted value of overall mortality and is adjusted for age.
   - Typically used in occupational epidemiology.
2. An intrinsic platelet disorder causing bleeding.
3. Platelets are large and thrombocytopenia may be present.
4. Platelets cannot attach to the subendothelium because they lack glycoprotein Ib for von willebrand factor (which normally mediates platelet adhesion).
5. Platelet transfusion (when necessary).
6. Platelet aggregation is normal in response to collagen, thrombin, ADP but not to risocetin.
7. Pulmonary edema.
8. Cardiac Tamponade. Do an echo first to confirm then insert a needle for pericardiocentecis then make a pericardial window.
9. Haloperidol.
10. Put the patient on low molecular weight heparin for prophylaxis.
11. Get a vaginal ultrasound. You need to make sure that the endometrial thickness is less than 4mm. If it's thicker than 4mm, you need to get an endometrial biopsy.

Day 3 Questions 4/5

Topic: antiretroviral (HIV) therapy.
1. When do you initiate the treatment?
2. what drugs?
3. what about Efavirenz?

Topic: treatment of chronic Hepatitis.
4. What is the treatment of chronic Hep B?
5. what is the treatment of chronic Hep C?

Topic: osteoporosis.
6. What is the number 1 risk factor for developing osteoporosis?
7. How do the laboratory results look?
8. How do you diagnose osteoporosis?
9. How do you manage a patient with osteoporosis?

◊

10. Weakness, pancytopenia, patient over 55 years old with massive splenomegaly. What kind of leukemia?
11. Weakness, pancytopenia, patient over 60 years old with hypogranular neutrophils… and the questions mentions neutrophils with a bilobed nucleus (pelger-huet) cells. What's the diagnosis?

Day 3 Answers 4/5

1. Either when the CD4 count is less than 350/µL or if the viral load is more than 55,000 copies.
2. You give a combo of 2 nucleosides (the vudine family or didanosine, DDC or tenofovirn) and one or more protease inhibitor (the Navir family··· i.e Indinavir)
3. A non-nucleoside reverse transcriptase inhibitor··· An alternative to the protease inhibitors.
4. Interfiron -or- Lamivudine.
5. Interferon –and- Ribavirin.
6. Being a white female.
7. all the labs are normal!
8. Diagnose osteoporosis by Dexa scan.
- If the T-score is >2.5 = osteoporosis.
- If the T-score is 1-2.5 = osteopenia, not osteoperosis.
9. You should give Calcium supplements with Vitamin D and put the patient on an exercise program if they are at the osteopenia stage. If you do the Dexa scan and you see that it is Osteoporosis, you need to add a Bisphosphonate to the treatment.
10. Hairy cell leukemia; a cancer of B-lymphocytes. Treat with Cladribine.
11. Myelodysplastic syndrome; a stem cell disorder (the bone marrow is hypercellular). Treatment is only supportive. Patients have an increased risk of infection due to the decrease in lymphocytes.

Day 3   Questions 5/5

Topic: Glanzman's thrombocytopenia.
1. What is it?
2. Would ristocetin work to aggregate the platelets?
3. How do you treat it?

Topic: Cushing's
4. How do you screen for Cushing's syndrome?
5. How do you treat Cushing's?

◊

6. How can you differentiate between acute stress disorder and post-traumatic stress disorder?

Topic: bleeding problems.
7. Prolonged PTT and normal PT:
8. Prolonged PT and normal PTT:
9. What if both PT and PTT are prolonged?
10. How can you tell if it is a factor 13 deficiency?

Day 3　　　　　　　Answers 5/5

1. A platelet disorder causing bleeding.
   - The count and the shape (morphology) of the platelets are normal. (as compared to the large platelets in Bernard-soulier)
   - The problem is due to a missing glycoprotein IIbIIIa.
2. Yes, in contrast to Bernard-Soulier.
3. Transfusion of platelets if necessary.
4. 24 hour urine level of free cortisol.
5. if not a surgical candidate, you can put the patient on Keto-conazol.
6. Timing is all what makes a difference! Acute stress disorder lasts only 4 weeks or less as compared to post-traumatic stress disorder which is more severe and lasts longer than 4 weeks, sometimes years.
7. If with no bleeding: Factor 12 deficiency.
   - With mild or rare bleeds: factor 11 deficiency (usually patient is an Ashkenazi Jew descendant).
   - With frequent or severe bleeding: factor 8 or factor 9 deficiency.
8. Factor 7 deficiency or early vitamin K deficiency… or look for use of Coumadin.
9. Defect in the common pathway (DIC, Liver disease, Vitamin K deficiency)
10. The clot solubility in 5M urea would be abnormal.

The USMLE Step Booster

1 DAY BREAK

Please take this time to review the material of the last 3 days.

SB

Day 4   Questions 1/5

Topic: Red Eyes

1. Presentation of a bacterial infection?
2. Presentation of a viral infection?
3. When do you suspect glaucoma in a patient with red eyes?
4. What about uveitis?

Topic: red eyes in a baby (ophthalmia neonatorum)

5. What is the cause if it is presented on first 24 to 48 hours after birth?
6. What if it occured between day 2 and day 5?
7. What if it started 7 to 14 days afterbirth (within the 2nd week of life)?

Day 4 — Answers 1/5

1. One red eye with a discharge (patients have crusting in the morning). Painless! Resolves spontaneously but sometimes needs to be treated with topical antibiotics.

2. Can be bilateral. Patients usually have a history of recent or current upper respiratory tract infection. Clear watery tears (no crusting). Most commonly cause by adenovirus so no treatment is necessary.

3. Glaucoma is extremely painful and there's always an abnormal pupil response to light (remember, if you suspect glaucoma, you need to get an ophthalmology consult).

4. Uveitis is also painful but the patients have decreased visual acuity and photophobia (if you suspect, do a slit light examination).

5. Chemical conjunctivitis due to silver nitrate drops.

6. N. Gonorrhea could be the cause at this time. Get a gram stain looking for gram negative intracellular diplococci. You can treat by giving one dose of Ceftriaxone.

7. Then you should think of Chlamydia trachomatis! So you need to get a Giemsa stain of scraped epithelial cells from the conjunctivae; intracytoplasmic inclusions will be present if the cause is actually Chlamydia. Treat with Erythromycin.

Day 4   Questions 2/5

Topic: Autoimmune Hemolytic Anemia... the difference.

1. Warm antibody induced:
2. Cold antibody induced:

Topic: Jaundice in a newborn

3. Would a bilirubin level of >5 mg/dl be physiologic jaundice if seen on first day of life?
4. What about a level of >12 mg/dl after the first 24 hours of life?

Topic: pneumonia in a newborn

5. What is the most common cause?
6. How do you treat it?
7. What about Chlamydia as the cause?
8. How would you treat that?

Day 4   Answers 2/5

1. Positive IgG or IgG + C3
- Drug induced, lymphoma or leukemia.
- Tx: steroids, splenectomy and if refractory, cyclophosphamide or cyclosporine.
2. Negative IgG and only positive for C3.
- Mycoplasma, Mononucleosis or Waldenstrom's.
- Tx: no steroids or splenectomy needed··· only cyclophosphamide or chlorambucil.
3. No! Bili >5 mg/dl on day 1 is pathologic!
4. This can't be physiologic either because physiologic jaundice would never give a bili level >12 mg/dl.
5. Group B strep, E.Coli and Listeria.
6. Ampicillin (to cover Listeria) plus an Aminoglycoside or Ampicillin plus Cefotaxime.
7. Chlamydia takes about 4 weeks to be the cause of pneumonia, so if the newborn presented with eye discharge and red eyes and then about 4 weeks later started having pneumonia, you should be thinking of Chlamydia as a possible cause.
8. Oral Erythromycin.

Day 4　　　　Questions 3/5

1. The normal level of PSA is 0 to 4 ng/ml, what if you get a patient with a level of 5 ng/ml on a screening test?

     Topic: postpartum endometritis.

2. What is the most important risk factor?
3. How do you treat it?
4. What about Metronidazole?

     Topic: Interstitial lung disease.

5. What is the x-ray findings of interstitial lung disease (restrictive)?
6. How would the pulmonary function test look like in such diseases?
7. What can you do to get a definitive diagnoses?
8. What is the only way a patient can have a restrictive pattern on pulmonary function testing (as above) but with a normal DLCo?

     Topic: viral meningitis.

9. What is the most common cause of viral meningitis in children?
10. What about in adults?

Day 4 — Answers 3/5

1. You need to send the patient to a urologist for a prostate biopsy!
2. Route of delivery! The problem occurs after only 3% of vaginal deliveries as compared to up to 30% of C-sections.
3. Clindamycin and Gentamycin.
4. Metronidazole is contraindicated in breast feeding patients.
5. Interstitial infiltrates with peripheral honey combs.
6. Decreased lung volume.
- Decreased DLCo
- Increased or normal FEV1/FVC.
7. You can get a transbronchial biopsy.
8. If there's extra pulmonary restriction such as kyphosis, obesity or a neuromuscular disease (basically anything that would prevent the chest to move up for the lungs to fully expand).
9. Arbovirus or Enterovirus. If the question mentions a rural setting, then you should think of Arbovirus because it is transmitted to the children through animal vectors.
10. The most common cause in adults is Herpes Simplex.

Day 4   Questions 4/5

Topic: Doc, I forgot my name!

1. Alzheimer's Disease:
2. Multiinfarct dementia:
3. Normal pressure hydrocephalus:
4. Brain Tumor:
5. Drugs:

◊

6. What is the initial treatment of pulmonary edema?
7. What is the treatment of Pelvic Inflammatory Disease?
8. How do you manage Rhabdomyolysis?

Day 4   Answers 4/5

1. Most common cause of dementia.
   - Not to be suspected in young patients (easy to rule out if patient is younger than 65 years old).
   - Slowly progressive, takes years to manifest.
   - Diagnosed by excluding all other possible causes.
2. Second most common cause of dementia.
   - Unlike Alzheimer's, deterioration in function is sudden but over a period of many years.
   - You can easily diagnose it by seeing the after effect on MRI or CT.
3. If the patient had dementia, ataxia and urinary incontinence, you should definitely suspect it.
4. Focal neurological findings would be present. Diagnosed by MRI or CT.
5. Suspect in acute settings especially in a young patient with an unknown history. A urine toxicology screen can solve the dilemma.
6. Oxygen, Morphine and a loop diuretic (usually IV furosemide).
7. IV cefotaxim (or ceftriaxone) + IV doxycycline.
8. Immediate adequate IV hydration followed by alkalinization of the urine.

Day 4 — Questions 5/5

1. What do you know about pseudohypoparathyroidism?

   Topic: CHF.
2. Treatment of CHF?
3. What about nitropruside and Enalaprilat?

   Topic: drug rehab.
4. How do you initiate the patient's recovery from drug addiction?
5. What about maintenance of drug abstinence?

Day 4 — Answers 5/5

1. PTH is there but the problem is that the organs are resistant to it, so it might as well not be there!
   - Low calcium levels and increased phosphate levels.
   - Short stature, mental retardation and short 4th and 5th metacarpals.
2. Put on oxygen mask.
   - Lower the pre-load:
   - Diuretics (furosemide: doubling the dose every 20 minute until you get sufficient urine output)
   - Nitrates.
   - Morphine.
   - If the patient is still short of breath, use positive inotropes (dobutamine or milrinone).
   - All your CHF patients have to leave the hospital with ACE inhibitors and beta blockers! DON'T EVER FORGET!
3. Those are only used in the ICU; they can be used to lower the after-load.
4. Initiation: drug rehab programs and in patient facilities.
5. Group therapy programs (i.e. Alcoholics Anonymous).

Day 5 — Questions 1/5

1. What are the lab findings in SIADH?
    Topic: lesions to the parietal lobe.
2. What symptoms would a patient have if they suffered from a lesion to the non-dominant Parietal lobe?
3. What about a lesion to the dominant parietal lobe?

◊

4. How do you manage a stable patient with ventricular tachycardia?
5. If the patient has a left ventricular abnormality or if the ejection fraction is less than 40%.
6. If you have a diabetic patient with low HCO$_3$ and low pH but a normal anion gap (basically a normal anion gap metabolic acidosis in a diabetic patient), what could this be due to?

Day 5   Answers 1/5

1. Low plasma osmolality, elevated urine osmolality and high urine sodium concentration.
2. Constructional (drawing, copying…etc.) and dressing apraxia. A way to remember: think of all the famous painters we all know that they all dress very oddly and most famous paintings are just a bunch of scribbles. Think of them as the non-dominant painters.
3. They can't do math (Gerstman's syndrome), finger agnosia, agraphia and not knowing their right from their left.
4. Give oxygen first and secure an IV access.
- Amiodarone or Lidocaine.
- Lidocaine 0.5mg/kg every 5 minutes but no more than a total of 3 mg/kg.
- Procainamide (max total of 1gm)
5. When is Amiodarone preferred to Lidocaine in the management of V-tachy in a stable patient?
6. This could be due to either RTA (type I or type IV) or a GI cause of acidosis.
- First thing to do is to calculate the urine anion gap:
  (Na+K)-CL. Normal gap is 0 to -50.
- If the urine anion gap is >0 then it is a renal problem (RTA)
- If the urine anion gap is < -50 then it could be due to a GI cause of acidosis such as loss of HCO_3 in diarrhea.

Day 5　　　　Questions 2/5

1. What is the difference between RTA type I and RTA type IV?
2. How does hypothyroidism cause infertility?
3. What is the best test to screen for acromegaly?
4. How do you confirm the diagnosis?
5. How do you treat myxedema coma?
6. If your hypothyroid patient became pregnant, how do you adjust the dose of levothyroxine?
7. How can you tell the difference between thyroiditis and factitious hyperthyroidism?
8. What antibiotic is associated with an increased risk of having a seizure!?

Day 5   Answers 2/5

1. RTA type I: distal renal tubular acidosis; NH3 is not excreted properly.
- RTA type IV: due to hypoaldosteronism (hyperkalemia).
2. Hypothyroidism --> increased TRH --> increased TSH (to try and stimulate the thyroid to fix the hypo problem).
- Along with TRH, prolactin production is also increased (since they both come from the anterior pituitary).
- Since prolactin has a negative effect on GnRH, it causes amenorrhea (remember that because post-partum breast feeding mother have amenorrhea for about 3 months because of high prolactin levels… it's the same thing but prolactin is increases via a different mechanism).
3. look for increased levels of Insulin like growth factor-1.
4. Get the oral glucose suppression test. Then get an MRI to locate the pituitary tumor. Remember, treat all pituitary tumors other than prolactinomas with surgery (transphenoidal).
5. IV hydrocortisone and IV levothyroxine.
6. You should increase the dose of synthroid by 30%
7. Radioactive Iodine uptake by the thyroid is decreased in both BUT thyroglobulin is low only in factitious hyperthyroidism.
8. Imipenem.

Day 5   Questions 3/5

Topic: Pain patients.

1. Describe Polymyalgia Rheumatica:
2. Fibromyalgia:
3. Polymyocitis:

◊

4. If you have a cystic fibrosis patient who's younger than 8 years and you decide to give the patient his/her first Influenza A vaccine, how would you do it?

5. If a child is 7 years old and you do not know the vaccination history, would you give Hib (H. Influenza type b) vaccination to this child?

Day 5 — Answers 3/5

1. Pain and stiffness of the shoulder and pelvic girdle.
- Difficulty in standing from a seated position (same as in polymyocitis) or lifting arm above the head (but unlike polymyocitis, polymialgia rheumatica patients do not have subjective weakness).
- Look for anemia and increased ESR
- Treat with low-dose prednisone.

2. Muscle pain, weakness and fatgue but no inflammation.
- Eather depression, anxiety or irritable bowel syndrome are almost always assocíted with this.
- Remember: you need to have 11/18 trigger points.
- Treat with TCA, physical therapy and stress reduction techniques.

3. These patients can eventually end up with difficulty in breathing or swallowing since it affects all types of muscle.
- look for an increase in creatinin, CPK, aldolase.
- Treat with high dose cortico-steroids.
- If you see a heliotrope rash: dermatomyocitis.

4. When younger than 8 years, give 2 doses of the vaccine 4 weeks apart.

5. No, you only need to give it to children younger than the age of 5 years.

Day 5 — Questions 4/5

1. Treatment of Kawasaki?
2. Treatment of Dermatitis Herpetiformis?
3. Treatment of Acne?
4. How do you diagnose Pulmonary Embolism?
5. How do you diagnose Deep Venous Thrombosis?
6. Treatment of Mycoplasma or Legionella Pneumonia?
7. What is the side effects of Ethambutol?
8. Describe Autoimmune Hepatitis:
9. What is the side effects of interferon in Treatment of Hepatitis C?
10. Treatment of community acquired pneumonia?

Day 5   Answers 4/5

1. Aspirin + IV Immunoglobulin.
2. Dapsone.
3. if comedones with minimal inflammation: Topical retinoids
- if severe with no inflammation: Oral Isoretinoin
- If inflamed: Antibiotics
4. 1st V/Q scan, if inconclusive then you have to order venous ultrasonography for DVT or CT angiogram. If both are negative, then do Pulmonary angiography which is considered "Gold Standard" but invasive.
5. Test of choice for suspected DVT: Compression Ultrasonography
- For recurrent DVT, study of choice: Impedence Plethysmography
6. Erythromycin.
7. Optic Neuritis.
8. Biopsy: portal inflammation with lobular damage "bridging necrosis"
- Prevalent in young females
- Increased Antinuclear and Anti-smooth Muscle Antibodies.
9. Depression.
10. In patient: Levofloxacin or gatifloxacin.
- Out patient: Azithromycin or doxycycline.

Day 5　　　　Questions 5/5

1. Most common cause of a mobile cavity mass in the lung, which presents with hemoptysis?
2. Two modalities that can decrease mortality in patients with COPD?
3. Describe Variable Deceleration:
4. What are the different types of Incontinence and their treatments?
5. What are the different types of Tocolytic Agents and their side effects?
6. What is the best Contraception after Pregnancy?
7. What do you think of when you see a painless lesion with punched out base and rolled edges, VDRL is negative and inguinal lympadenopathy is present:
8. Fever, headache and confusion. What diagnosis should you think of?
9. If RBC's and WBC are increased in lumbar puncture?

Day 5   Answers 5/5

1. Aspergilloma.
2. Home Oxygen and smoking sessation.
3. Sudden 15 beats/min decrease in fetal heart rate.
- Lasting for 25 seconds.
- Due to umbilical cord compression.
- Treatment: Mask Oxygen and chage maternal position. If it fails and pattern persists do: Scalp pH.
4. Motor Urge incontinence: Oxybutynin
- Overflow incontinence: Bethanicol + alpha-blocker
- Stress incontinence: Kegel exercises + Urethroprexy
5. Respiratory depression + Muscle Weakness: Mg Toxicity Treatment: Calcium gluconate.
- Tachycardia + hypotension + myocardial depression: Calcium Channel Blocker.
- Hypokalemia + Hypoglycemia: Beta Agonist
- Oligohydramnois + PDA closure in utero: Indomethacin
6. Progestin only because does not affect breast milk. Use after 3 months if the mother is breast feeding or 3 weeks if not.
7. Syphillis: remember, serologic testing is not reliable early in course of disease. Diagnose made by dark field microscopy.
8. Encephalitis.
9. Herpes. Treat it with Acyclovir.

Day 6   Questions 1/5

1. Describe Endometriosis:
2. Describe Adenomyosis:
3. Pain of 1st stage of Labor:
4. Pain of 2nd stage of Labor:
5. What is painless early cervix dilatation and how do you treat it?
6. What is the Secretin Stimulation test?
7. Describe the D-Xylose Test and its use:

                Topic: Aortic transection.

8. What is the study of choice:
9. If it shows a widen mediastinum, what is the next step?
10. What if mediastinum is not widened, what is the next step?

Day 6   Answers 1/5

1. Dysmenorrhea, dyspareunia and dyschezia
- Physical exam shows tender adnexal mass and firm nodularity located in the broad ligaments, the uterosacral ligament or in the cul-de-sac.
2. Endometrial glands in the uterine muscle
- Symptoms: dysmenorrhea, menorrhea
- Physical exam: enlarged and symmetrical uterus.
3. T10-12 (use narcotics).
4. S2-S4: "perineum pain" (don't use narcotics).
5. Incompetent cervix. Treatment: cerclage = suturing the cervix.
6. Measure the baseline gastrin level then inject secretin (IV)
- In normal individuals, Gastrin decreases (antacid effect)
- In Zollinger-Ellison Syndrome, Gastrin increases.
7. Have the patient drink D-xylose and then measure it in their blood stream. If positive, it means absorption is normal. Can be done either in 2-hr blood test or 5-hr urine test.
8. X-Ray.
9. Perform an Aortogram.
10. Perform a spiral CT or Transesophageal echo.

Day 6   Questions 2/5

1. Treatment of Status Epilepticus:
2. Most common cause of sinusitis, otitis, bronchitis pneumonia? Treatment?
3. Prophylaxis of infective endocarditis in GU or GI procedures?
4. What if the patient allergic to penicillin?
5. Describe Eustachian Tube Dysfunction:
6. Describe Fetal scalp sampling:
7. What biomarkers do you follow for recurrent cardiac re-infarction?
8. Describe Isolated Diastolic Dysfunction:

Day 6   Answers 2/5

1. 1st: Lorazepam/Diazepam.
- 2nd: Phenytoin/Phosphenytoin.
- 3rd: Phenobarbital.
- 4th: Midazolam/Propofol
2. #1 Strep Pneumonia
- #2 H. Influenza
- #3 N. Meningitidis
- Treatment: Amoxicillin
3. Ampicillin + gentamycin
4. Vancomycin + gentamycin
5. Follows an upper respiratory infection, presents as aural fullness and an audible pop in the ear whole swallowing and yawning. It causes conductive hearing loss.
6. pH> 7.25, expectant management
- pH between 7.20 and 7.25 then the FSS should be repeated in 15 to 20 minutes
- pH< 7.20, deliver baby
7. CK-MB
8. Normal Size heart
- Normal ejection fraction
- Normal left ventricle end diastolic volume
- S4 gallop present
- Treatment: Beta-blocker

Day 6   Questions 3/5

Topic: Fever after Surgery!

1. If around 102:
2. If around 105:
3. Day 1 post surgery:
4. Day 3 post surgery:
5. Day 5 post surgery:
6. Day 7 post surgery:
7. Two weeks post surgery:

◊

8. Describe Trichotillomania?
9. Treatment of Impetigo?
10. Treatment of Gonorrhea and Chlamydia? when do you treat what?
11. What is Mittelschmerz?
12. Prophylaxis for PCP in HIV Pt < 200 CD4 count?
13. Why choose Azithromycin over Clarithromycin for M. Avium prophylaxis?
14. What is the most effective agent used for the treatment and prevention of seizures in eclampsia?
15. What test is 99% sensitive and specific when checking for pneumonia caused by Legionella?

Day 6　　　　　Answers 3/5

1. Atelectasis.
2. If shortly after anesthesia: Malignant Hyperthermia.
- Shortly after instrumentation: Bacteremia
3. Atelectasis.
4. Urinary Tract Infection.
5. Thromophlebitis.
6. Wound Infection.
7. Deep Abscess (Subphrenic, Pelvic).
8. Pulling ones own hair.
9. Oral Erythromycin or Topical Mupirocin.
10. When treating Gonorrhea, you must treat Chlamydia at the same time but not true for reverse cases.
11. Abdominal pain in a young female between her menstrual cycles with a benign clinical exam.
12. TMP-Sulfa: If rash develops, give Dapsone instead.
- If rash develops but patient is allergic to dapsone or is G6PD, give Atovaquone!
13. Because Azythromycin is only once a week!
14. MgSO4.
15. Urinary Antibody Antigen.

Day 6 Questions 4/5

1. If a patient came to you with Pneumonia, hyponatremia and diarrhea (or any gastrointestinal symptoms), what should you think of? how do you treat it?
2. Treatment of mild PID (pelvic inflammatory disease)?
3. Treatment of Severe PID?
4. Painless nodules or papules which rapidly ulcerates, does not have rolled edges or punched out base. Characterized by irregular borders and has a beefy red granular base, what do you think it is?
5. What about Granuloma Venereum?

   Topic: Treatment of Multiple Sclerosis.
6. Acutely:
7. To slow progression of remitting episodes:
8. Treatment of spasticity:
9. Treatment of fatigue:
10. Treatment of urinary incontinence:
11. Treatment of urinary retention:

Day 6 — Answers 4/5

1. Legionella, Treatment: Erythromycin.
2. Outpatient: Single dose of Ceftriaxone (IM) + Doxycycline (PO) or Azithromycin.
3. In Patient: clindamycin + Gentamycin
- or Cefotetan (IV) or cefoxitin (IV) + Doxycycline
4. Granuloma inguinale.
5. Painless, shallow and associated with systemic nonspecific symptoms. The inguinal adenopathy is inflammatory and does not appear at the same time as the ulcer.
6. Corticosteroids.
7. B-Interferon or glatirimic acid.
8. Baclofen.
9. Amantadine.
10. Oxybutamine.
11. Bethanechol.

Day 6 — Questions 5/5

1. "Speckled pattern" on echo is specific for what? What cardiac problems does this cause?
2. Difference between contrictive and restrictive cardiomyopathy?
3. EKG findings in pericardial tamponade?
4. Fever, headache plus focal lesions with a CT showing ring or contrast enhancing lesions, what's the diagnoses?
5. Describe Savage Syndrome:
6. Describe DUB (dysfunctional Uterine Bleeding):

Day 6　　　　　Answers 5/5

1. Amyloidosis, causes a retrictive cardiomyopathy.
2. Restrictive: X-ray may show calcified pericardium and CT scan usually shows increased thickness of the pericardium.
- Constrictive: Thickness of the myocardium will be normal.
3. Electrical alternans. Other findings include sinus tachycardia and low voltage QRS complexes.
4. Can be cancer or infection. If patient is HIV negative, get a biopsy.
- If patient is HIV positive, first treat with pyrithamine + sulfadiazine and repeat CT within few days. If CT shows improvement then you can say that toxoplasmosis is the culprit. If CT shows same as before Treatment then you must biopsy.
5. Ovarian resistance syndrome.
- A congenital failure of the ovaries to respond to FSH and LH secondary to receptor defect.
- Patients have primary amenorrhea and absent sexual characteristics.
6. Most common cause of abnormal uterine bleeding
- Diagnosed by exclusion
- Most common cause in adolescent women is anovulation.
- IV estrogen is the drug of choice for uncontrolled bleeding

The USMLE Step Booster

1 DAY BREAK

Please take this time to review the material of the last 3 days.

SB

Day 7   Questions 1/5

1. Hypertension before 20 weeks gestation:
2. What is Pseudocyesis?
3. What is puerperal fever?
4. What causes puerperal fever?
5. What do you know about endometritis?
6. Study of choice for an acute episode of diverticulitis?
7. What is the Treatment of recurrent bed wetting?
8. Steeple sign on X-Ray is pathognomonic for what disease?
9. Thumb print sign on X-Ray is pathognomonic for what disease?
10. Describe Cephalohematoma:
11. Describe Caput Succedanum:

Day 7	Answers 1/5

1. Could be either chronic hypertension or hydatidform mole.
2. Imaginary pregnancy, requires psych evaluation.
3. Defined as increase in temp > 38C for more then 2 consecutive days in the first 10 days post partum.
4. Most common cause is endometritis but it may be caused by UTI or atelectasis.
5. Typically occurs on the 2nd or 3rd day post partum.
- Risk factors: prolonged labor, prolonged and premature rupture of membranes, unusual removal of the placenta and repeated pelvic examinations.
- It's a polymicrobial infection. 70% anaerobics.
- Treatment: Clindamycin with either an Aminoglycoside or Ampicillin.
6. CT scan.
7. 1st line: Desmopression.
- 2nd line: Imipramine.
8. Croup (Parainfluenza).
9. Epiglottitis (H. Influenza).
10. One bone of baby's head is swollen (does not cross suture lines).
11. More than one bone involved (crosses suture lines).

Day 7   Questions 2/5

1. Most specific test for iron deficiency Anemia?
2. Most specific test for Sideroblastic Anemia?
3. Most Specific test for Thalessemia?
4. Most common cause of toxic megacolon?
5. What do you get to test for Lactose intolerance?

Topic: Pneumonia on X-Ray.

6. If you see lobar infiltrates?
7. If you see bilateral interstitial infiltrates?
8. What is the most common cause of lobar penumonia in any age group?

◊

9. Describe the Progesterone withdrawal Test:
10. Treatment of Hypertension with stable angina?
11. What if they cannot tolerate it or it is contraindicated?
12. When do you use the Kleihauer-Betke test?

Day 7   Answers 2/5

1. Bone marrow for stainable iron.
2. Prussian blue stain for ringed sideroblasts.
3. Hemoglobin Electrophoresis.
4. Ulcerative colitis.
- Treatment: IV steroids, nasogastric decompression and fluid management (X-ray shows dilatation)
5. Positive hydrogen breath test.
- Positive clinic test for reducing substances.
- increased stool osmotic gap.
6. Regular bacterial pneumonia: Strep, Staph Hameophalus, Klebsiella.
7. Mycoplasma, Chlamydia, Legionella, Pneumocystitis or Viral.
8. Strep Pneumonia.
9. Intramuscular injection of progesterone for 5 to 10 days to a women who stopped menstruating for unknown reasons.
- If she has withdrawal bleeding, that means that she has normal endogenous estrogen production.
- If not then it is menopause or Ahserman's Syndrome.
10. Beta blockers.
11. Ca-Antagonists.
12. To check if fetus blood is present in mother's system.

Day 7   Questions 3/5

1. what is the effect of BCG on PPD testing?
2. what is Couvelaire Uterus?
3. Placenta Previa?
4. Vasa Previa?

                 Topic: Triple Markers.

5. Diagnose ↑ AFP, ↔ Estriol, ↔ HCG
6. Diagnose ↓ AFP, ↓ Estriol, ↑ HCG
7. Diagnose ↓ AFP, ↓ Estriol, ↓ HCG

                 ◊

8. Describe Malignant Otitis Externa?
9. Describe Pericarditis on EKG:

Day 7 Answers 3/5

1. Does not change treatment recommendations because BCG does not cause a skin reaction to go over 10 cm.
2. "bruised uterus" severe abruption, painful bleeding.
3. Fetus in transverse lie position, painless bleeding.
4. Amniotomy, painless bleeding, bradycardic fetus, requires immediate c-section.
5. Most likely dating error so the next step is to do Sonogram, Amniocentesis for fluid AFP and Karotype!
- Can be: Ventral Wall defect, Twins, Placental Bleeding, fetal renal disease, Neural Tube defect (follow up with Acetyl-cholinesterone levels).
6. Most Likely Dating Error or Down's Syndrome.
7. Most Likely Dating Error or Trisomy 13.
8. Caused by pseudomona aeruginosa, usually in diabetic patients or immune compromised patients.
- Foul smelling ear discharge (greenish), granulation in ear canal and involves cranial nervs VII, IX, and XII.
- Diagnose: CT scan.
- Treatment with Ciprofloxicin and debrivement.
9. ST segment elevation in all leads, PR segment depression.

Day 7 — Questions 4/5

1. What is cholesteatoma?
2. Treatment of early disseminated and late Lyme disease?
3. Treatment of early-localized Lyme disease? what if the patient is pregnant?
4. What is Riluzole?
5. Treatment of Fibromyalgia?
6. What age should you administer pneumococcal vaccine?
7. What is the most serious complication of bronchiectasis?

Topic: HIV Prophylaxis.

8. <200 CD4: What organism should you use prophylaxis against and which antibiotic to use?
9. <50 CD4: What organism should you use prophylaxis against and which antibiotic to use?
10. When should HIV treatment begin?

◊

11. Describe Constrictive Pericarditis on physical examination.

Day 7   Answers 4/5

1. Epithelial cyst that contains desquamated keratin. Patient presents with chronic ear discharge and granulation tissue that are unresponsive to antibiotic Treatment.
2. Ceftriaxone or cefotaxime.
3. Oral Doxycycline, but if pregnant use Amoxicillin.
4. A glutamate inhibitor approved for use in patients with ALS.
5. Amitriptyline and cyclobenzaprine.
6. At age 65 not any younger unless it's a special case.
7. Hemoptysis (Xray: "Tramtrack") diagnose by High resolution CT.
8. PCP, Treatment with TMP-SMX unless allergic, treat with Dapsone.
9. M. Avium, treat with azithromycin.
10. T cell <350 or viral load >55,000.
11. JVD, ascites, clear lungs, positive knock.
- Caused by any chronic dz that would cause fibrosis.
- Diagnose: Xray or CT
- Treatment: Remove the pericardium

Day 7   Questions 5/5

1. Treatment of Herpes Simplex and Varicella?
2. Treatment of Herpes simplex, Varicella and CMV?
3. What Drugs can cause ototoxicity?
4. Patient with recurrent attacks of positional vertigo and tinnitus with sensorineuronal hearing loss, Diagnose?

      Topic: Feeding Problems in Children.

5. Feeding problems plus hypoplastic lungs, think of?
6. Dropping platelet count + Premature baby, think of?
7. Cystic Fibrosis, Common major complication?
8. Progessive jaundice at age 8 weeks, think of?

◊

9. Treatment of Partial Seizures?
10. Treatment of Generalized Seizures?
11. Treatment of myoclonic/atonic seizures?
12. Treatment of Unidentifiable seizures?
13. Treatment of Absence Seizures?
14. Describe Diamond-Black Fan Syndrome?
15. Describe Fanconi's Anemia?

Day 7   Answers 5/5

1. Acyclovir, Valcyclovir, Fancyclovir.
2. Gancyclovir, Foscarnet, Cidofovir.
3. Furosemide.
- Aminoglycosides.
- Vancomycin.
- Quinine.
- Chloroquine.
4. Menier's Disease.
5. Diaphragmatic hernia.
6. Necrotizing enterocolitis.
7. Meconium Ileus (diagnose with gastrographic enema).
8. Biliary atresia, perform HIDA scan after one week of phenobarbital.
9. Carbamazepine or Phenytoin (2nd line" Valproic Acid)
10. Valproic Acid or Lamotrigine.
11. Valproic Acid.
12. Valproic Acid.
13. Ethosuxamide.
14. A child with macrocytic anemia, low reticulocytic count and increased apoptosis.
- Treatment: Corticosteroids.
15. Absent thumb, café-au-lait, pancytopenia, macrocytosis, microcephalic, positive horse-shoe kidneys.

Day 8   Questions 1/5

Topic: when you suspect meningitis.

1. Which patients require a CT scan of the head before lumbar puncture?
2. What do you do before CT if needed?
3. From the Lumbar puncture, what is the most specific test for meningitis?
4. From the Lumbar puncture, what is the most sensitive test for meningitis?
5. Next best initial test on CSF is?

Topic: Vaginitis (do speculum and microscopic examination!)

6. Presentation of Candida vaginitis:
7. Presentation of Hemophilus vaginitis:
8. Presentation of Trichomonas vaginitis:

◊

9. Treatment of catatonic schizophrenia?
10. What is the presenation and treatment of Tricyclic Antidepressant (TCA) intoxication?
11. Management of post-term pregnancy?

Day 8　　　　　　　Answers 1/5

1. Any patient who has focal findings, papilledema or altered mental status.
2. Always give ceftriaxone before CT.
3. Culture.
4. Protein.
5. Next best test is the cell count.
6. "Sticky" discharge (adhesive to vaginal lateral wall), microscopic exam shows "hyphae", pH~4
- Treatment: Antifungal.
7. "Fishy Odor" especially with KOH (whiff test), microscopic exam shows "clue cells", pH~5
- Treatment: Metronidazole.
8. "Frothy" discharge, Microscopic exam shows flagellated organisms, pH~6
- Treatment: Metronidazole (Note: In 1st trimester pregnancy, treat with clindamycin!)
9. Lorazepam.
10. Seizures, hypotension, prolonged QRS
- Treatment: Sodium-Bicarbonate
11. Anything >42 weeks must do non-stress test and biophysical profile twice weekly.

Day 8   Questions 2/5

1. Effect of pregnancy on Bun/Cr?
2. How do you Treatment active phase arrest?
3. Identify the deficient enzyme and physical exam findings in Nieman-Picks disease?
4. Describe the deficient enzyme and physical exam findings in Tay-sach's disease?
5. Gaucher's disease, identify the deficient enzyme and classic presentation:
6. Krabbe's disease, identify the deficient enzyme and classic presentation:
7. Mucopolysaccharidoses, identify the deficient enzyme and classic presentation:

        Toipc: Juvenile Rheumatoid Arthritis.

8. Describe Type I:
9. Describe Type II:
10. Describe Type III:

Day 8 Answers 2/5

1. Bun/Cr in pregnancy is 40-50% of the normal.
2. Treat with C-section (when there is strong contractions with no dilation).
3. Spingolipidosis due to deficiency in sphingomyelinase.
- Hypotonia, hepato-splenomegaly, cervical lympadenopathy, protruding abdomen, cherry red macula.
4. Spingolipidosis due to deficiency in hexamindase A.
- Hyperacusis, mental retardation, seizures, cherry red macula, no hepatosplenomegaly or cervical lympadenopathy.
5. Deficiency in glucocerebrocidase.
- Hepato-splenomegaly, anemia, thromocytopenia (no cherry red macula).
6. Deficiency in galactocerebrocidase.
- hyperacusis, irratibility and seizures.
7. Characterized by coarse facial features, hydrocephalus and umbilical hernia.
8. ANA (+), ↑ girls, Iridocyclitis (always do eye exam).
9. HLA-B27, ↑ in boys, ↑ in ankylosing spondylitis.
10. Systemic symptoms: fever followed by rash then involvement of the joints.

Day 8　　　　Questions 3/5

1. What is the most common lung cancer associated with asbestos exposure?
2. What lung cancer is almost exclusively associated with asbestos exposure but not the most common malignancy after exposure?
3. Describe Acute bronchopulmonary aspergillosis (ABPA):

Topic: I just can't hold it!

4. Describe Detrusor Instability (Urge incontinence):
5. Describe Overflow Incontinence:
6. Describe Stress Incontinence:

◊

7. What's Osgood Schlatter?
8. What's nursemaid elbow?

Day 8   Answers 3/5

1. Bronchogenic carcinoma.
2. Malignant mesothelioma.
3. A trascient recurrent pulmonary infiltrate, pheripheral eosinophilia, asthma and immediate wheal and flare reaction to Aspergillus fumigatus and presence of Abs in the serum against Aspirgillus. ↑ IgE in serum.
- ↑ in Cystic Fibrosis patients.
- On X-Ray: "Cluster of grapes"
- Treatment: Oral Prednisolone.
4. Detrusor instability, bladder irritation from a neoplasm and interstitial cystitis resulting in urge incontinence: sudden and frequent loss of moderate to large amounts of urine often accompanied by nocturia and frequency.
5. Can be caused by diabetic neuropathy: loss of small amounts of urine from an over distended bladder and increased residual volume.
6. A loss of small of urine simultaneously occurring with activities that increase intra-abdominal pressure, along with physical examination demonstrating pelvic floor weakness.
7. Swelling of tibial tubercle, common in physically active pubertal children.
8. Radial head subluxation, Treatment: gentle supination.

Day 8   Questions 4/5

1. Describe Osteosarcoma:
2. Describe superficial thrombophlebitis and its Treatment:
3. What is the most common side effect of an epidural block?

    Topic: HIV patient with pain in the esophagus:

4. Differential Diagnoses?
5. Work up?
6. Treatment of each differential?

◊

7. What is the most likely reason why a patient on dialysis would have worsening hypertension?

Day 8   Answers 4/5

1. Most common cause of bone cancer in children, ↑ risk of bilateral retinoblastoma.
- On X-Ray: Bone sclerosis (+)
2. Dull pain in the region of the affected vein.
- Erythema, induration and tenderness along the vein, no edema or deep calf tenderness, Not DVT.
- Treatment: NSAIDS
3. Hypotension.
- Treatment: Ephedrine and left lateral position with hydration.
4. If thrush is present, then it's most likely Candida.
- If thrush is absent, think of CMV, HSV, HIV.
5. CMV and HSV look the same, must do a biopsy to differentiate.
- CMV: Cytomegallic cells
- HSV: Ground class cells
- If neither one, then by exclusion it is HIV induced ulcer (Treatment: Prednisone).
6. Candida: Fluconazole.
- CMV: Gancyclovir / Foscarnit.
- HSV: Acyclovir, famcyclovir, Valcyclovir.
7. Most likely due to erythropoetin.

Day 8   Questions 5/5

Topic: Child with sickle cell presenting with fever:

1. If the child is on penicillin prophylaxis with temperature < 40 C, what do you do?
2. What if the child is not on prophylaxis or appears toxic with temperature > 40 C?

◊

3. What do you prescribe for migraine prophylaxis?
4. What do you use in treating acute attacks of migraine?

Topic: Kallman's syndrome.

5. What is Kallman's syndrome?
6. What labs do you get?
7. How do you treat it?

◊

8. What is Amaurosis Fugax?
9. What does that tell you if your patient has it?
10. So what should you do if your patient has it?
11. What test to order if you suspect organophosphate poisoning?

Day 8 Answers 5/5

1. Send blood cultures, ceftriaxone IV or IM and observe for a few hours in ER then you can follow up as an outpatient.
2. Blood cultures, Ceftriaxone IV and hospitalization.
3. For prophylaxis give Methysergide.
4. Sumatriptan or naratriptan.
5. The combination of Anosmia and Hypogonadotroin.
6. Decreased LH, FSH and testosterone.
7. Give testosterone and for the patients that want to have children, give GnRH.
8. A sudden transient blindness in one eye only.
9. It tells you that the patient has carotid atherosclerosis, so check for a carotid bruit.
10. You should get a carotid Doppler to evaluate the patient's need for a carotid endarterectomy.
11. RBC cholinesterase level. It would be low if there's poisoning because prganophosphates are cholinesterase inhibitors.

Day 9   Questions 1/5

1. What is the cause of restless leg syndrome?
2. How do you treat it?
3. How do you treat Aplastic anemia?
    Topic: correcting hyponatremia.
4. If the patient's hyponatremia is a chronic problem?
5. If the patient becomes worse over a few days but they say that his hyponatremia is a chronic problem?
6. What if it is an acute problem and the patient is having mental status changes by the minute!?

◊

7. Treatment of choice for catatonic Schizophrenia?
8. Treatment of Bipolar patients with renal dysfunction?
9. Treatment of Social Phobia?

Day 9   Answers 1/5

1. Iron deficiency.
2. ropinirole and pramipexole.
3. If the problem is mild, you only give supportive treatment.
   - If neutrophills <500, platelets <20,00, reticulocytes <1% and the patient is younger than 50 years and you have an HLA matching sibling then you need to refer for a bone marrow transplant··· if no donor or HLA match, you can give Cyclosporin.
4. Treat only by restricting fluids to 1-2 L per day.
5. Give NaCL (normal saline) and a diuretic.
6. Give hypertonic saline (3%) and try to correct the sodium halfway within the first 24 hours.
7. Lorazepam (Benzodiazepine).
8. Not Lithium! Treat with Valproate or Carbamazepine.
9. Assertive training with SSRI drugs.

Day 9   Questions 2/5

1. Incontinence, Dementia, Ataxia, Diagnose?

    Topic: Adenomatous Polyps.
2. Which type of polyp has a higher risk to become cancerous, Sessile or Stalked?
3. Put the following in order from most likley to least likely to become malignant: Tubular, Tubulorvillous and Villous:
4. Does size increase the risk of malignancy?

    ◊

5. How do you calculate Birth-Rate?
6. Fertility-Rate?
7. Fetal Mortality Rate?
8. Maternal Mortality Rate?
9. Why do patients with Zollinger-Ellison Disease experience malabsorption?
10. What side effect can you get from excessive use of oxytocin?

Day 9   Answers 2/5

1. Normal Pressure Hydrocephalus.
2. Sessile Polyps.
3. Most Likely to become malignant: Villous Adenomous.
- 2nd: Tubulovillous.
- 3rd: Tubular.
4. Negligable risk if < 1.5 cm
- 2-10% risk if 1.5 to 2.5 cm
- 10% risk if polyps > 2.5 cm
5. # of Live Births / 1,000 Total Population.
6. # of Live Births / 1,000 women 15-45 years of age
7. # of Fetal Deaths / 1,000 Total Births
8. # of Maternal Deaths / 100,000 Pregnancies
9. Due to the inactivation of pancreatic enzymes by increased production of stomach acid.
10. May cause water retention, water intoxication, Hyponatremia and seizures.

Day 9   Questions 3/5

1. If a patient has a mild penicillin allergy (rash) what do you use?
2. What if the allergy is severe?
3. Describe the Tanner Stages:
4. How to diagnose Kidney Stones:

Day 9   Answers 3/5

1. Use first generation cephalosporins: 5% cross reactivity to penicillin
- Cefazolin (IV)
- Cephalexin
- Cephradine
- Cephadroxil

2. Clindamycin: (cover: Staph, Strep, Anaerobes)
- Macrolides: Erythromycin, Clarithromycin, Azithromycin for life threatening infections.
- Vancomycin, Synercid, Linezolid: Used for gram positive infections with life threatening allergy to penicillin and for MRSA.

3. I: Breast development but no pubic hair
- II. Breast budding or thelarche, small amount of pubic hair growth spurt usually follows
- III. Breast development but no contour separation, dark hair
- IV. Breast development with contour separation, more pubic hair
- V. Complete breast development with no contour separation of areola (pubic hair is of adult quality and distribution)

4. 1st: do UA, check for hematuria
- 2nd: Confirm diagnose with non-contrast spiral CT
- If pt is pregnant, sono is test of choice.

Day 9　　　Questions 4/5

Topic: Female Infertility (remember, 5-10% Unexplained)
1. Peritoneal factor:
2. Ovulatory factor:
3. Cervical factor:

◊

4. Pre-renal Vs. Acute Tubular Necrosis?
5. Treatment of Otitis media and Sinusitis?
6. Treatment of Pharyngitis?
7. Treatment of Lyme disease, Chlamydia, Rickettsia?
8. Treatment of Urinary Tract Infection in pregnancy?

Day 9 — Answers 4/5

1. Most common cause includes endometriosis and peritoneal adhesions. Laparoscopy is procedure of choice for diagnosis and Treatment. Mild cases respond to GnRH agonist such as Danazol and Medroxyprogesterone.

2. Can be screened for by checking basal body temperature (assesses the duration of luteal phase) and mid-luteal level of progesterone (assesses the level of luteal phase)

3. Structural anamolies and abnormal mucus production, check musuc exam and postcoital tests.

4. Pre-renal: BUN/Cr > 20:1
   Urine Na Low < 20
   Urine Osm High > 500

   ATN: BUN/Cr 10:1
   Urine Na High > 40
   Urine Osm Low < 350

5. Amoxicillin.
6. Penicillin.
7. Doxycycline.
8. Nitrofurantoin.

Day 9　　　　　Questions 5/5

1. Sensitive test to Diagnose Clostridium Difficile (next step in management)?
2. Pathognomonic of Chron's disease as compared to Ulcerative colitis?
3. Patient with positive PPD and positive Chest x-ray, do you start treating?
4. Describe TFT in Pregnancy (even if she had hypothyroidism before):
5. Most common type of incontinence caused by epidural anesthesia?
6. Triad of renal failure, microangiopathic hemolytic anemia and thrombocytopenia is pathognomonic for what disease?
7. What is the most common side effect of cimitidine?
8. Mechanism of action of methochlorpromide? side effects?

Day 9 — Answers 5/5

1. Cytotoxin assay in the stool.
2. Non-caseating granulomas.
3. No, first get sputum smear and cultures to see if they have active disease. If they do, don't give INH alone it will increase the resistance.
4. ↑ total T4, normal free T4, ↑ TBG, normal TSH,
   - ↑ TBG due to TSH action of BHCG + ↑ Estrogen
5. Overflow Incontinence. Best treated with intermittent catherization.
6. HUS! Treat with exchange transfusion, plasmapharesis with administration of FFP.
7. Gynecomastia in men and confusion in the elderly.
   - Note: Inhibits P450 system, potential to ↑/↓ other medications.
8. Inhibits dopamine at the LES, dopamine relaxes the LES so low dopamine = ↑ LES tightening which decreases reflux.
   - Also, methoclorpromide ↑ gastric motility
   - Side effects: inhibiting dopamine in CNS: slowness, tiredness, depression, moodiness, etc.

The USMLE Step Booster

1 DAY BREAK

Please take this time to review the material of the last 3 days.

SB

Day 10 — Questions 1/5

1. Describe Reactive NST:
2. Describe Biophysical Profile:
3. Describe Contraction Stress Test (CST):
4. Describe Whipple's Disease:

Day 10   Answers 1/5

1. Two or more accelerations in 20 minutes (>15 BPM for > 15 seconds).
2. Includes: NST, Amniotic fluid by sono, fetal breathing movements, fetal gross movements, extremity tone.
- Each profile is worth 2 points with a minimum of 0 points and maximum of 10 points.
- 8 to 10: Good results, repeat weekly
- 4 to 6: If > 36 weeks, deliver.
-      If <36 weeks, do a CST or repeat BPP in 4-8 hrs.
- 0 to 2: Poor results, deliver immediately
3. Three contractions in ten minutes.
- Absence of late decelerations is a negative test = good results, repeat weekly.
- Positive late decelerations = Must deliver immediately.
4. A multi-systemic illness. Weight lost, abdominal pain, diarrhea, malabsorption, distention, flatulence and steatorrhea.
- Patient may experience migratory polyarthritis, chronic cough, CHF or valvular regurgitation.
- Later stages can include dementia and myoclonis.
- Biopsy shows PAS-positive material in the lamina propria of the small intestine.
- Electron Microscopy is the most specific test for confirmation.

Day 10   Questions 2/5

1. Deficiencies in complement factors C6, C7, C8, think of?
2. Deficiency in C1q, think of?
3. Deficiencies in C1rs, C1s, C2, C4, think of?
4. Deficiency in either C3 or C5, think of?
5. How do you calculate osmolality?
6. As serum Glucose increases, what happens to Sodium?
7. Treatment of osteoporosis?
8. When should steroids be used to Treatment SLE?
9. What is Lichen Simplex Chronicus?
10. Describe Mitral regurgitation:

Day 10 — Answers 2/5

1. Neisseria meningitis.
2. SLE-like syndrome.
3. SLE-like syndrome and glomerulonephritis.
4. Pyogenic infections.
5. 2(serum Na) + (Glu/18) + (BUN/2.8)
- Or, if BUN is normal, use: 2(Na) + 10
6. For every 100 units of increased Glucose, Sodium drops by 1.6 units.
7. Calcium, Vitamin D and Alendronate (a bisphosphonate: works by inhibiting osteoclasts as well as by binding to hydroxyapatite to decrease bone resorption.
8. Anytime there is renal involvement, thrombocytopenia, hemolytic anemia, CNS involvement, and or SLE pericarditis.
9. AKA: localized scratch dermatitis and neurodermatitis
- Extreme itching causing a patch of dry skin, seeding, hyperpigmented and thickened skin
- A ring of discrete brownish papular can be seen at the periphery of the lesion
10. A pansystolic murmur heard best at the apex and radiating to the axilla
- Treatment: just like CHF: Digoxin, enalapril and thiazides.

Day 10   Questions 3/5

1. List some acute manifestations of ESRD (which determines need for dialysis):
2. What is the most common side effect of clozapine?

   Topic: Abortion.
3. 1st trimester bleeding + Normal sonogram + closed cervix?
4. Abnormal sonogram + closed internal cervical Os?
5. Abnormal sonogram + opened internal cervical Os + No products of conception passed?
6. Abnormal sonogram + opened Os + tissue present on sonogram?
7. Abnormal sono + opened Os + empty uterus?

   ◊

8. Arrest disorder of dilation + prominent ischial spines, what is the diagnosis?
9. Define arrest disorder of dilation:
10. Treatment of Alzheimer's Dz?
11. Describe Ramzy Hunt Syndrome:

Day 10　　　　Answers 3/5

1. Hyperkalemmia, metabolic acidosis, fluid overload, pericarditis, encephalopathy.
2. Agranulocytosis, this is why it is only used in refractory psychosis even though it is the most effective atypical antipsychotic.
3. Threatened Abortion.
4. Missed Abortion
- Management: Scheduled D&C or Observation
5. Inevitable abortion.
- Management: Emergency D&C.
6. Incomplete abortion.
7. No tissue left = complete abortion.
8. Midpelvic contractions.
9. Cervical dilation being the same for > 2 hrs OR descent that has not progressed for > 1 hr.
10. Donepezil.
11. Facial nerve paralysis caused by Herpes Zoster.
- Patient presents with lesions of the external ear with or without tympanic membrane involvement, vertigo, tinnitis and deafness.

Day 10 Questions 4/5

1. What is the treatment of endometriosis?
2. Urine dipstick is positive but no RBC's are found microscopically, explain:
3. Treatment of Parkinson's in non-functional patients?
4. Treatment of Parkinson's in functional patients?
5. What is ropinirole?
6. What drug arrests the progression of Parkinson's disease?
7. What glomerulonephritis is it that follows a virla infection?
8. Post-strep glomerulonephritis, next step?
9. Urine eosinophils, most common diagnoses?
10. What test is first used in the diagnose of Zollinger-Ellison Syndrome?

Day 10   Answers 4/5

1. Lupron (leuprodie) for 4-6 months
- Birth control pills
- Provera
- Danazol (side effects: hirsutism)
2. Myoglobinuria.
3. Carbidopa/Levodopa.
4. If > 60 year old, treat with Amantadine
- If < 60 year old, treat with Anticholinergics (ie. Trihexiphenidate).
5. A drug that simulates dopamine receptors in the brain, decreases tremor and rigidity, usually added to carbidopa/levodopa.
6. selegeline.
7. IgA glomerulonephropathy: Berger's disease.
8. Get ASO titers.
9. Allergic Interstitial Nephritis (due to drugs).
- Stop the drug that is causing it (commonly, cephalosporins).
10. Serum gastrin concentration (fasting). Measurement of the gastric pH is a must to exclusde the possibility of secondary hypergastrinemia due to achlorhydria.
- Patients with non-diagnostic fasting serum gastrin levels should have a secretin stimulation test done.

Day 10   Questions 5/5

1. Classic primary ovarian failure in Turner Syndrome?
2. Testicular torsion, and treatment?
3. Epididimytis, management and treatment?
4. Treatment of Multiple Myeloma:
5. Treatment of CLL?
6. Describe Bartter's Syndrome:

Day 10   Answers 5/5

1. FSH elevation greater then LH elevation.
- Why? because there is no estrogen and feedback to negatively inhibit FSH and LH
2. No diagnostic studies needed, straight to surgery!!
- Orchiopexy may also be done on the other testes.
3. Do sonogram first to rule out testicular torsion.
- Fever, pyuria, sexual active age, testes in normal position.
- Treatment: Antibiotics
4. Melphanon and Prednisone.
5. Chlorambucil + Prednisone OR Fludarabine.
6. Defective Sodium and Cloride reabsorption in the thick ascending limb of Henle.
- Hypovolemia results which activates the renin-angiotensin-aldosterone system.
- RAAS causes an ↑ in Potassium and H ion secretion which leads to ↓ Potassium and metabolic alkalosis.
- Early in life as polyuria, polydipsia and growth/mental retardation (which can present later in life)
- Normal blood pressure!
- Rule out use of diuretics!

Day 11   Questions 1/5

1. First line Treatment for Idiopathic benign Intracranial hypertension?
2. Drug used as prophylaxis in Toxoplasmosis?
3. Drug used to treat Toxoplasmosis?

   Topic: Valvular Dysfunctions (Exam: Transthoracic-Echo).
4. Describe the physical findings and treatment in Mitral Stenosis (MS): (As MS worsens, opening snap moves closer to S2)
5. Describe the physical findings and treatment in Aortic Stenosis (AS): (Systolic Crescendo-Descrendo)
6. Describe the physical findings and treatment in Mitral Regurgitation (MR): (Holosystolic/Pan-systolic)
7. Describe the physical findings and treatment in Aortic Regurgitation (AR):
8. Idiopathic Hypertrophy Subaortic Stenosis (IHSS), Mitral Valve Prolapse (MVP):

Day 11   Answers 1/5

1. Acetazolamide.
2. For prophylaxis, use TMP-SMX.
3. For Treatment, use sulfadiazine and Pyrimethamine.
4. Presents with Hemoptysis, Dysphagia, Emboli, atrial fibrilation and hoarsness.
- Increased in pregnancy.
- Treatment: 1st: Diuretics, if it fails, treat with balloon.
- Ace-Inhibitors and Digoxin does not work in treating MS.
5. Presents with Syncopy, CHF and angina.
- Treatment: Must Replace the valve.
6. Presents like CHF (so treat it like one!).
- Treatment: Preload Reduction, Positive Inotrope, Afterload reduction.
7. Presents like CHF.
- Treatment: Preload Reduction, Positive Inotrope, Afterload reduction
8. Treatment: Never give diuretics! treatwith a B-blocker.
- Systolic mumur at left lower sternal border.
- ↑ Blood in heart = ↓ Murmur
- ↓ Blood in heart = ↑ Murmur

Day 11   Questions 2/5

Topic: Teratogens!!!

1. Effect of Alcohol?
2. Effect of DES?
3. Effect of Isoretinoin?
4. Effect of Lithium?
5. Effect of Tetracycline?
6. Effect of Trimethadione (anticonvulsant)?
7. Effect of Warfarin?
8. Effect of Valproic Acid?
9. Effect of Dilantin?
10. Effect of Streptomycin?

Topic: EKG's.

11. Describe a Transmural Ischemia on EKG:
12. Describe a Transmural Infarct on EKG:
13. Describe Subendocardial Ischemia on EKG:
14. Describe Subendocardial Infarct on EKG:
15. Describe Ventricular Aneurysm on EKG:

Day 11　　　　　Answers 2/5

1. Midface Hypoplasia + Long Philtrum.
2. Mullerian Anomalies.
3. Microtia (small ears) + CNS and Cardiac Anaomalies.
4. Ebstein Anomaly (R heart Failure).
5. Discolored Teeth.
6. Facial Dysmorphism.
7. Stippled Epiphesis (optic atrophy).
8. Neural Tube defect (Spina Bifida).
9. Craniofacial Desmorphism + Nail Hypoplasia.
10. 8th Nerve Damage.
11. ST segment elevation, absence of Q waves and negative cardiac enzymes.
12. ST segment elevation followed by development of Q waves and ↑ cardiac enzymes.
13. ST segment depression.
14. ST segment depression not followed by Q waves and elevation of cardiac enzymes.
15. Usually follow an acute MI, has persistent ST segment elevation.

Day 11   Questions 3/5

Topic: Tests in Anemia.

1. What is the most specific and sensitive test in Autoimmine Anemia? Treatment?
2. What is the most specific and sensitive test in Glucose-6-phosphate-Deficiency? Best Treatment?
3. What is the most specific and sensitive test in PNH? Treatment?
4. What is the most specific and sensitive test in Hemolytic-Uremic Syndrome (HUS)? Treatment?
5. What is the most specific and sensitive test in Hereditary Spherocytosis? Treatment?

◊

6. Big baby, skin lesions, huge placenta, thrombocytopenia, resulting in death, Diagnose?

Day 11   Answers 3/5

1. Test: Coomb's Test.
- Treatment: Initially treat with steroids.
- In life threatening hemolysis, give IV Immunoglobulin.
- If recurrent, spleenectomy.
2. Test: G6PD level (invalid if cells are young, must wait).
- Treatment: Avoid oxidants!
3. Test: Sugar-water and Ham's test.
- Treatment: Steroids.
4. Diagnosis is made based on physical exam. Renal Failure + Thrombocytopenia + Hemolysis (No Specific/Sensitive tests).
- Treatment: Initially spontaneous resolution, however if life threatening, treat with plasmapharesis
- Note: If fever + neurologic problems are also present then the diagnosis is TTP.
- Treat TTP with Plasmapharesis!
5. Test: Spherocytes on smear and osmotic fragility test
- Treatment: Splenectomy.
6. Syphilis: Transmission to baby through placenta
- Treatment: Benzathine Penicillin even if mother is allergic.

Day 11 Questions 4/5

Topic: Naughty kids!

1. Describe Anti-social children:
2. How is that different than Conduct Disorder?
3. What about Oppositional Defiant Disorder?

◊

4. Which semester is most detrimental to the fetus in being infected with Toxoplasmosis?
5. Which trimester is the most common time of transmitting Toxoplasmosis to fetus?
6. Which trimester is most detrimental to the fetus in being infected with Rubella?
7. Which trimester is the most common time of transmitting Rubella?
8. How does a baby present if infected with Toxoplasmosis?
9. If female patient is IgG negative for Rubella, next step in management?
10. What is the screening test for gestational diabetes?

Day 11   Answers 4/5

1. Violates the right of others, older than 18 year old.
- For example, lie, steal, destroy property, set fires, cruel to animals and people.
2. Violates the rights of others but is younger that 18 years of age.
- For example, lie, steal, destroy property, set fires, cruel to animals and people
3. Disobedient and argumentative; hostile but does not seriously violates the rights of others.
4. 1st trimester.
5. 3rd trimester.
6. 3rd trimester.
7. 1st trimester (so it's the exact opposite to toxo).
8. Chorioretinitis + Intracranial calcifications + symmetrical IUGR.
9. Give vaccine (live) and inform her not to get pregnant for 3 months (but if she accidentally did, it's not so bad).
10. Done at 24-28 weeks
- 1 hour, 50 gm glucose test, should be < 140
- Follow up test: 3 hr, 100 gm (after 8-10 hrs of fasting) read at 1,2,3 hrs. 2 abnormal readings is confirmatory.

Day 11   Questions 5/5

Topic: Hernias.

1. Describe indirect hernias?
2. What about a direct hernia?
3. Describe the femoral hernia:

◊

4. What is the difference between Mania and Hypomania?
5. Describe Brief psychotic disorder:

Day 11 — Answers 5/5

1. MC type in both men and females
- Herniation of the gut through the internal ring of the inguinal canal
- When the finger's placed in the external ring of the inguinal canal, the examiner feels the herniation on the tip of his finger.
2. Protrudes directly through the abdominal wall.
- Age and obesity increase the risk.
- Hernias are felt along the lateral aspect of the physician's finger when its placed in the external inguinal ring.
3. More common in females, particulary during and after pregnancy
- Not located in Hesselbach's triangle.
- Hernia travels beneath the inguinal ligament and down the femoral canal (↑ risk of strangulation).
4. Mania: At least one week of DIGFAST.
- Hypomania: Less than one week (typically less then 4 days) of DIGFAST symptoms.
5. Delusions, hallucination, disorganized speech/behavior
- At least one day but less than one months and should completely resolve before you diagnose it.

Day 12 Questions 1/5

1. What causes epigastric pain when a patient is suffering from severe pre-eclampsia?
2. Describe Dysthymic disorder:
3. Describe Bipolar II disorder:
4. What is the Sestamibi scan?
5. What type of lung cancer causes an increase in Ca?
6. What do you need to know about Sarcoidosis?
7. When do you see Calcium Oxalate Crystals?
8. Describe Nesidioblastosis:
9. Describe a baby with galactokinase deficiency:
10. Describe a baby with galactose-1-phosphate uridyl transferase deficiency:

Day 12　　　　　Answers 1/5

1. Stretching of the liver capsule "Glisson's"
2. Depressed mood most of the time for 2 years or more.
- Poor appetite, over eating, insomnia/hypersomnia, low-self esteem, low energy, hopelessness and or poor concentration.
3. Episodes of major depression and hypomania
- Manic episodes are NOT seen
4. Used to locate a PTH tumor if one is suspected.
5. Squamous cell carcinoma due to PTHrh.
6. Sarcoidosis can produce an increase in Calcium.
- Bilateral hilar adenopathy + Erythema Nodosum.
- Diagnosed by biopsy.
7. Rectangular, enveloped shaped crystals in the urine, seen in patients with ethylene glycol (anti-freeze) poisoning.
8. A baby with high insulin levels and low blood glucose.
9. Only cataract is seen, otherwise asymptomatic.
10. Failure to thrive, bilateral cataract, jaundice and hypoglycemia.
- Increased risk of E. coli neonatal sepsis.

Day 12 Questions 2/5

Topic: Vertigo!

1. How does Central (pure) vertigo present?
2. Describe Peripheral (mixed) vertigo:
3. What are the differentials for peripheral vertigo?

◊

4. Define Abortion:
5. Define Antipartum-Death:
6. Define Intrapartum-Death:
7. Define Fetal-Death:
8. Define Neonatal-Death:
9. Define Infant Death:
10. Define Maternal Death:

Day 12 Answers 2/5

1. Vertical nystagmus.
- Does not suppress with fixation.
- Multi-directional.
- Test: MRI
2. Horizontal nystagmus.
- Suppresses with fixation.
- Unidirectional.
3. Menier's Disease: Vertigo, hearing loss, tinnitus, Treatment: Diuretics and low sodium diet.
- Benign Postural Paroxysmal Vertigo, Treatment: Movement exercises.
- Labrynthitis: Usually follows an upper respiratory tract infection. Treatment: Meclazine or diazepam.
4. Pregnancy loss prior to 20 weeks gestation.
5. Death of baby during anytime from 20 weeks gestation to onset of labor.
6. Death of baby during anytime from labor to birth.
7. Death of fetus at any time from 20 weeks of gestation to birth.
8. Death of neonate at any time from 20 weeks of gestation to 28 days after birth.
9. Death of Infant defined as any time from birth to 1 yr of age.
10. Death of mother during pregnancy or 90 days after delivery.

Day 12   Questions 3/5

1. Buspirone is the drug of choice for which disorder?
2. SSRIs are the drugs of choice for which disorder?
3. What is the mechanism of action of TCAs? what are they used for?
4. What does the clinical diagnose of ALS require?
5. Describe Huntington's Disease:
6. Genetics of Cystic Fibrosis:

Day 12   Answers 3/5

1. Treatment of Anxiety disorders in which abuse or sedation is a concern!
2. 1st line Treatment for depression
- Takes 3 to 4 weeks to work
- Side effects: Major sexual Dysfunction.
- some common SSRIs: Fluoxetine, Sertraline, Paroxetine.
3. Imipramine, Amytriptyline, Desipramine are a few examples
- Blocks neuroephinephrine and serotonin reuptake
- 2nd line for depression and also used in neurologic pain
- Easy to overdose and die from prolonged QT
4. Requires signs of disease in three extremities or two extremities plus the face.
5. An Autosomal dominant disease (Multiple CAG repeats).
- Movement and mental dysfunction starting in middle age.
- Genetic analysis detects chromosome 4 abnormality.
- CT/MRI: Caudate nucleus and putamen atrophy.
- Fatal in < 20 years from diagnosis.
6. Deletion of a three base pair encoding for phenylalanine in the CFTR gene on chromosome 7.

Day 12　　　Questions 4/5

1. What is Zoledronic Acid?
2. How do you treat TB meningitis?
3. What should you know about Cryptococcus meningitis:
4. What GI problem is most commonly associated with Henoch-Schonlein?
5. Treatment of uncomplicated cystitis and pneumocystic pneumonia?
6. Rheumatoid arthritis + Splenomegaly + neutropenia, diagnose?
7. Rheumatoid arthritis + Lung problems, diagnose?
8. What diagnosis should you think of when you see a patient with Parkinsonism experiencing orthostatic hypotension, impotence, incontinence or other autonomic symptoms?
9. What is Riley-day Syndrome?
10. Hepatic cyst with small cyst inside of it, diagnose?
11. Prevention of pertussis if a family member has the disease?

Day 12   Answers 4/5

1. It is a bisphosphonate.
- Recommended in all patients who have metastatic breast cancer and lytic bone disease.
2. Steroids.
3. Affected patients are usually HIV (+) with CD4 < 50
- First diagnostic test: India Ink.
- Cryptococcal antigen testing is very sensitive and specific.
- Treatment: 1st Amphoteracin then fluconazole for life.
4. Intussusception .
5. Trimethaprim sulfamethoxazole.
6. Felty's Syndrome.
7. Caplan's Syndrome.
8. Shy-Dragger.
9. Orthostatic hypotension + autonomic dysfunction seen mostly in Ashkenazi Jewish children. Autosomal recessive.
10. Echinococcus.
11. All close contacts should be given erythromycin for 14 days regardless of age, immunization history or presence and absence of symptoms (just give it to everyone).

Day 12   Questions 5/5

1. Treatment of recurrent bed wetting?
2. What is the classic triad of congenital rubella?

           Topic: Hips in Kids.

3. Refusal to move hip after febrile illness, diagnose?
4. What about pain in bone after febrile illness, next step in management?
5. Painless limping turns painful in a patient who is around 6 years of age?
6. Hip pain in early teen male, diagnose?
7. What hip problem is common in newborns?

           ◊

           Topic: Sounds like Autisitc.

8. Describe Asperger disorder:
9. Describe Rett Syndrome:

Day 12   Answers 5/5

1. Use the alarm first (Pavlovian conditioning).
- if drugs are needed, use Desmopression (DDAVP) as first line and Imipramine as second line.
2. Sensorineural deafness, cardiac malformation (PDA, ASD) and Cataracts.
3. Septic hip, emergency drain needed with general anesthesia.
4. Osteomyelitis: no x-ray, do bone scan and treat with antibiotics.
5. Avascular necrosis of femoral head.
6. Slipped femoral epiphysis.
7. Development dysplasia, order sono not x-ray.
8. Patients are more communicative than Autistics and appear more socially aware.
- Do not have language impairments as in autistics.
- Do have social impairments, repetitive behavior and obsessional interests.
9. "Rett girls", x-linked dominant.
- Development starts normal until the child is around 1 year of age. Then the language and motor skills regress and microcephaly develops. (remember: Hand wringing and sighing are characteristics).

The USMLE Step Booster

1 DAY BREAK

Please take this time to review the material of the last 3 days.

SB

Day 13   Questions 1/5

1. How is Pyelonephritis distinguished from Cystitis? 1st step in management? Best test? Treatments?

   Topic: Chest Pain.

2. What should you think of when the patient has chest Pain that changes with position?
3. What about chest Pain that changes with palpation?
4. What is the first test you should always do in all patients who complain of chest pain?
5. Describe the EKG leads assigned for: Inferior-wall, Anterior-wall, and Lateral-Wall.
6. How do you treat a stable patient with V-Tachycardia?
7. What if the patient was unstable?
8. Treatment of Atrial Fibrillation in a patient with Wolf-Parkinson-White Syndrome?

Day 13 — Answers 1/5

1. 1st test: Urinalysis.
- Best Test: Culture.
- Cystitis: Low-grade fever, suprapubic pain. Treatment: Sulfa or Cipro for 3 days (follow up culture if pregnant in 1 week)
- Pyelonephritis: High-grade fever, Flank pain. Treatment: Any gram-negative drug.
- Do ultrasound to determine cause of pyelonephritis and don't forget to admit the patient if she's pregnant.
2. Pericarditis.
3. Costochondritis.
4. EKG! (easy question but easily forgotten!)
5. Inferior wall: II, III, IVF.
- Anterior-wall: V2, V3, V4.
- Lateral-wall: I, IVL, V5, V6.
6. Lidocaine.
7. Shock! (defibrillation).
8. Procainamide or disopyramide (do not use Digoxin or any of the Ca-channel blockers!).

Day 13   Questions 2/5

1. Describe Condyloma Accuminata:
2. Describe Condyloma Lata:
3. What are the "TORCH" infections?
4. How can you diagnose Rocky Mountain Spotted Fever, Lyme, Syphillis and Mycoplasma?
5. What cardiac abnormality do polycystic kidney patients have?

Day 13   Answers 2/5

1. Valvular Papillomatosis
- Genitcal lesions caused by HPV 6 and 11
- Exophytic lesions with a raised papillomatous or spike surface and may grow into a large cauliflower-like shape
- Biopsy: Koilocytes.
2. 2nd stage of syphillis (think: Lata = later)
- It ulcerates and may present accompanied by a maculu-papular rash.
3. "TORCH"
- **T**oxoplasmosis:
- -Treatment for mother: Spiramycin or Pyrimethamine + Sulfonamide.
- -Treatment for baby: Pyrimethamine + Sulfa + Leukovorin for 6 months.
- **O**ther: syphillis, Varicella.
- **R**ubella: Diagnose by ordering IgM titers.
- **C**MV: Treat with Gancyclovir
- **H**erpes Simplex: Treat with Acyclovir.
4. Gram stain cannot be done, so you get Serologic Testing (IgM and IgG's).
5. 25% of these patients have mitral valve prolapse.

Day 13 — Questions 3/5

1. Transudate Vs. Exudate:

    Topic: Congenital Masses in the Neck!!

2. A mass located midline at the level of the thyroid gland, moves up when the patient protrudes the tongue. What could it be?

3. Along the line of the anterior edge of the Sternocleido-mastoid muscle between SCM and the pharynx (lateral surface of neck), Diagnose?

4. Supraclavicular, spongy in consistency located at the base of the neck, Diagnose?

Day 13　　　　Answers 3/5

1. Not easy to memorize but very important... sorry!

	Transudate	Exudate
Specific Gravity	<1.015	>1.015
Total Protein	<3.0	>3.0
Fluid/Serum Protein Ratio	<0.5	>0.5
Fluid/LDH Ratio	<0.6	>0.6
Fluid/Serum Glucose Ratio	>1.0	<1.0
LDH	<200	>200

2. Thyroglossal Duct Cyst (becomes painful if infected)
3. Brachial Cleft Cyst.
4. Cystic Hygroma (usually present by age three)

Day 13   Questions 4/5

1. What should you know about Zenker's (Pharyngoesophageal) Diverticulum?
2. Describe the Keith-wagener Classification:
3. Define Preterm labor: How, what, when and treatment?

Day 13   Answers 4/5

1. Develops as a balloon above the upper esophageal sphincter by herniating posteriorly between the fibers of cricopharyngeal muscle.
- Pathology: Motor dysfunction and incoordination putting pressure on the wall.
- Treatment: surgery (cricopharyngeal myotomy).
2. A system used to describe the severity of Hypertensive Retinopathy.
- Grade I: Arteriolar narrowing and tortuosity, increased light reflex and slight AV nipping.
- Grade II: Copper wiring, AV depression.
- Grade III: Cotton-wool exudates, flame shaped hemorrhages.
- Grade IV: Flame shaped hemorrhages, exudates and papillary edema.
3. Labor occurring before 37 weeks of gestation
- To diagnose, you need documented uterine contractions at 4 contractions / 20 minutes and documented cervical changes or cervical effacement of 80% or cervical dilation of 2 cm or more.
- Treatment: first only treat with bed rest and hydration (hydration decreases secretion of ADH and Oxytocin) If unsuccessful, give a Tocolytic.

Day 13 Questions 5/5

1. Treatment of any Life Threatening Fungal Infections (ie. Endocarditis, Meningitis, Fungemia)?
2. Treatment of Candida Infections?
3. Treatment of Onychomycosis (nail infection)?

   Topic: Anaerobes!

4. Treatment of oral anaerobes (above the diaphragm)? (can be an aspiration pneumonia)
5. Treatment of Abdominal anaerobes (below the diaphragm)?

   ◊

6. Describe the Staging of CLL:
7. What drug should be avoided in Treatment of Post-traumatic Stress Disorder?

Day 13 Answers 5/5

1. Amphotericin (Renal failure and RTA are the side effects)
2. Any "...azole" (ie. Fluconazole, Ketaconazole, Itraconazole)
3. Terbinafine, Itraconazole.
4. Clindamycin.
- Penicillin (any penicillin except Oxacillin,cloxacillin,Dicloxacillin,Nafacillin).
5. Metronidazole
- Imipenem, Meropenim.
- 2nd generation cephalosporin.
- Beta-lactam/beta-lactamase inhibitor combination.
6. Stage 0: Lymphocytosis (↑WBC).
- Stage 1: Lymadenopathy.
- Stage 2: Spenomegaly.
- Stage 3: Anemia.
- Stage 4: Thrombocytopenia.
7. Benzodiazepines.

Day 14   Questions 1/5

Topic: Infective endocarditis.

1. What pathogen should you think of when the patient has damaged valves?
2. What if the valves were prosthetic?
3. What if the patient is an IV drug abuser?

◊

4. What are the side effects of thiazide diuretics?
5. What is the treatment of Streptococcal pharyngitis?
6. What is the HELLP syndrome?
7. So how can you HELP these patients?
8. Describe Tetralogy of fallot:

Day 14 Answers 1/5

1. Streptococcus viridans.
2. Staphylococcus epidermis.
3. Staphylococcus aureus (from his skin).
4. Hyperglycemia, increased LDL, increased triglycerides, hyponatremia, hypokalemia and hy**per**calcemia (decreases Ca excretion in urine).
5. A single dose of intra muscular Benzathine Penicillin G or a 10 day course of oral Penicillin V. Use erythromycin if the patient is allergic to penicillin.
6. Hemolysis, Elevated enzymes, Low Platelets.
- A severe variant of pre-eclampsia.
- life-threatening.
7. Give IV MgSO$_4$.
- Induce labor.
- Lower the blood pressure.
- Maternal steroids.
8. Pulmonary Stenosis: causes a harsh systolic murmur at upper left sternal border.
- Ventricular septal defect (VSD) : causes cyanosis.
- Right ventricular hypertrophy.
- Aorta overriding the VSD.
- The most common cause of "blue babies"
- "Boot-like" appearance of heart on x-ray.

Day 14   Questions 2/5

1. Describe Primary Sclerosing Cholingitis (PSC):
2. Describe Primary Biliary Cirrhosis (PBC):
3. What is the worst prognostic sign of pre-eclampsia? What is the worst prognostic sign of pre-eclampsia?
4. What is the standard of care for threatened abortion?
5. What marker should be followed for bone formation?
6. What marker should be followed for bone resorption?
7. What are the characteristic signs and symptoms of Di-George's Syndrome?
8. Describe Steinert's disease (Myotonic Muscular Dystrophy):

Day 14 — Answers 2/5

1. Fibrosis of intra-hepatic ducts (can be visualized on ERCP)
   - Associated with Ulcerative Colitis
   - Cholestatis, Jaundice, Pruritis
2. Females : Males (9:1)
   - Same symptoms as PSC but (+) Anti-mitochondrial Antibody.
   - Biopsy needed for diagnosis.
3. Retinal Hemorrhage (due to retinal vasospasm).
4. Just reassurance and outpatient follow up.
5. Alkaline phosphatase.
6. Urinary N-telopeptide.
7. T-cell deficiency, hypocalcemia, truncus arteriosis, fish-mouth.
8. Autosomal Dominant
   - 2nd MC muscular dystrophy
   - Delayed muscle relaxation
   - All types of muscles involved
   - Thin cheeks and upper lip looks like inverted "V"

Day 14　　　Questions 3/5

1. Treatment of Hepatitis B?
2. Treatment of Hepatitis C?
3. What is a common side effect of Interferon?
4. What is the Legg-Perthes-Calvé disease?
5. What is the most common cause of a baby not being able to urinate during the first 24hrs of life?

Topic: Sad Mothers.

6. Describe maternity blues:
7. Describe postpartum depression:
8. Describe postpartum psychosis:

Day 14 Answers 3/5

1. Lamivudine or interferon.
2. Interferon and ribavirin in combination.
3. Depression.
4. Avascular necrosis of femoral head after trauma (most common reason). The patient is usually a child between 2 and 12 years of age. They usually have a limp and pain in the hip, groin or knee.
5. Most likely due to posterior urethral valves.
- 1st step in management: Empty the bladder immediately to preserve the renal function.
- 2nd: VCUG is the single best imaging modality to detect PUV (Important).
6. Feelings of sadness, dysphoria, sudden mood swings, tearfulness and dependence.
- Occurs in up to 80% of women.
- Due to the rapid change in women's hormonal levels and stress associated with maternity. Nothing major to worry about.
- the exact opposite of "baby pinks!" which can be a mild to severe form of mania.
7. Requires symptoms of major depression lasting longer than one week. Unlike maternity blues, this occurs more often in the months following child birth rather than immediately after delivery.
8. Requires the presence of hallucinations in addition to frequent suicidal or infanticidal ideation.

Day 14   Questions 4/5

1. What drugs are used in the treatment of Influenza?
2. What diagnosis should you think of when a mother brings her baby to your office complaining that he turns blue when feeding?
3. Describe what an arthrocentesis would show in a non-inflammatory, inflammatory and septic joint:
4. What is Amaurosis Fugax:
5. Define Todd's paralysis:
6. What diagnosis should you think of when you see Pneumatosis Intestinalis on abdominal film?
7. What diagnosis will show "White reflex" in a child's eye exam?
8. Aneuridia + Hemihypertrophy in a child. What is the diagnosis?
9. Describe an Omphalocele:
10. Describe Gastroschisis:

Day 14 — Answers 4/5

1. Zanamivir, osteltamivir.
2. Choanal atresia: closure of one or both posterior nasal cavities.
- Diagnoses is confirmed by ordering a CT scan with intra nasal contrast.
3. WBC count in joint fluid:
- Non-inflammatory: < 2,000
- Inflammatory: 5,000-50,000
- Septic > 75,000
4. Unilateral blindness in a hypertension patient which resolves on its own.
5. A post-ictal paralyisis, a condition that usually rapidly improves with restoration of motor function within 24 hrs of seizure. Sudden loss of consciousness is characteristic.
6. Necrotizing enterocolitis
- Treatment: Bowel rest and antibiotics.
7. Retinoblastoma.
8. Wilm's tumor.
9. Covered by a sac, intestines cannot be seen located at midline.
10. Not covered by sac, intestines are seen, is not located midline (usually located to the right of the umbilicus).
- Gastroschisis and Omphalocele are surgical emergencies.

Day 14   Questions 5/5

Topic: Variations in Fetal Heart Rates... what's the cause?

1. Early deceleration?
2. Variable deceleration?
3. Late deceleration?

◊

4. What is the difference between placenta previa and abruptio-placenta?
5. Which antihypertensive drug is known to cause depression as a side effects?
6. A positive Lachman maneuver or pivot shift test is specific for what?
7. The McMurray maneuver is specific for what?
8. Describe the Acid/Base in pregnancy:
9. What is the etiology of having metabolic acidosis in chronic renal disease?
10. Describe Post-ictal lactic acidosis:

Day 14 — Answers 5/5

1. Associated with head compression.
2. Associated with cord compression.
3. Due to uteroplacental insufficiency.
- Placenta Previa:
  - Lower then normal placenta attachment
  - Bright red bleeding
  - No contractions
- Abruptia Placenta:
  - Early separation of placenta.
  - Common in pre-eclampsia.
  - Dark red blood.
  - Contractions present.
4. Beta-blockers (Propanolol).
5. Injury to ACL.
6. To check for lateral and medial meniscal injuries.
7. Chronic respiratory alkalosis.
8. Impaired ammonia excretion.
9. It is transient and resolves without treatment withn 60 to 90 minutes. You should only observe and repeat labs after 24hrs).

Day 15   Questions 1/5

Topic: Breasts.

1. 18 year old with a rubbery mass, diagnose?
2. 20-40 year old with multiple lumps, tender, related to cycle, diagnose?
3. Bloody nipple discharge, diagnose?
4. Older lady, hard mass, skin dimpling, diagnose?

Topic: Meningitis by Age!

5. Most common cause in newborns?
6. Most common cause in 1 month–2 yrs old.
7. Most common cause in 2-18 yrs old:
8. Most common cause in 18-60 yrs old:
9. Most common cause in 60 yrs old and above:

◊

10. Treatment for an acute panic attack?
11. Long-term treatment of a panic attacks?
12. Treatment of Obsessive Compulsive Disorder?
13. What is the most safest and effective drug for the treatment of clinical depression in a patient with a history of stroke?
14. Describe Nihilistic delusions:
15. What is Canesthetic hallucinations?

Day 15 Answers 1/5

1. Fibroadenoma.
2. Fibrocystic disease.
3. Intraductal papillema.
4. 1st diagnosis to consider is cancer.
5. Group B strep > E. coli > Listeria > H. influenza.
6. S. pneumonia, N. meningitis > Group B strep, Listeria, H. influenza.
7. N. meningitis > S. Pneumonia > Listeria.
8. S. pneumonia > N. meningitis > Listeria.
9. S. pneumonia > Listeria > gram negative rods.
10. Benzodiazepine (esp. Alprazolam).
11. SSRIs.
12. SSRIs or clomipramine.
13. SSRIs (Sertraline).
14. False feeling that self or others or a significant aspect of the self (such as one's heart) does not exist.
15. False sensations of things occurring in or to the body (think of tactile hallucination).

Day 15   Questions 2/5

1. Nasal obstruction, nasopharyngeal mass and recurrent epistaxis in a young male is a triad for what disease?
2. What is Quinsy?
3. Describe the symptoms of retropharyngeal abscess:
4. What should you suspect when the patient has migratory thrombophlebitis with venous thrombosis?

                 Topic: Polyps.

5. Hyperplastic Polyps:
6. Hamartomatous Polyps:
7. Adenoma:
8. If you see a polyp when doing a sigmoidoscopy, what should you do?

Day 15   Answers 2/5

1. Juvenile Angiofibroma. If you suspect it, you need to get a CT of head and face.
2. Peritonsillar abscess: unilateral sore throat, neck pain and respiratory distress, increased in children.
3. Posterior pharyngeal edema, nuchal rigidity, cervical adenopathy and fever.
4. This could be symptoms of chronic DIC due to a cancer!
- Labs would show a slight increase in PT, a decrease in fibrinogen, and an increase in Fibrin split products
- Do: CT of chest, abdomen and pelvis to locate the cancer.
5. The most common of the non-neoplastic colorectal polyp. You don't need to do any further workup if you see one.
6. Includes juvenile polyp (a non-malignant lesion, removed to avoid its risk of bleeding) and peutz-jeghers polyp (also, a non-malignant lesion).
7. Most common type.
- found in up to 50% of the elerdy population.
- Usually asymptomatic but you need to be careful because it can be pre-malignant (even if only <1% become malignant).
8. You need to see the rest of the colon to see if there's more of it; Order a colonoscopy.

Day 15   Questions 3/5

1. How do you treat Prinzmetal's angina (also known as variant angina)?
2. A lesion to what part of the brain would give a patient a presentation of Hemi-neglect Syndrome:
3. What are the side effects of ovulation induction (gonadotropin injection)?
4. What are you worried about if a fetus was exposed to phenytoin in utero?
5. How do you manage a tubo-ovarian abscess?
6. What are the indications and the side effects of Bupropion?
7. What is Trazodone used for?
8. How do you treat exercise indused asthma?

Day 15 — Answers 3/5

1. Basically, it is a vasospasm, so you can treat it with Ca-channel blocker.
2. The right (non-dominant) parietal lobe.
3. Multiple gestations and Ovarian Hyperstimulation Syndrome
- Remember: it may cause ovarian torsion.
4. Neuroblastoma.
5. Usually treated with triple antibiotic cocktail including gentamycin, clindamycin and ampicillin. If that doesn't help within 24 hours, the abscess must be drained!
6. Used to treat depression with fatigue and difficulty in concentrating or the same patient is depressed and has attention deficit disorder.
- Side effects: Insomnia and weight loss (that makes sense).
7. Treatment of depression with significant insomnia (so the side effect is drowsiness).
8. Give the patient Cromolyn. Also, you can tell them to use a beta agonist inhaler before exercising.

Day 15   Questions 4/5

1. How do you manage Hyperkalemia?
2. Name some causes of hyperkalemia:
3. How do you treat Syphilis?
4. Treatment of Bacterial Meningitis?

Day 15 — Answers 4/5

1. If you see that the Potassium level affected the EKG, you need to treat with CaCl or Calcium gluconate.
- Otherwise, give glucose and insulin or Bicarbonate.
- Then remove Potassium from the body by sodium polystyrene (Kayexalate) or by dialysis if renal failure was the cause of the problem.

2. K-sparing diuretics, Hypoaldosteronism, Ace inhibitors, Renal failure, Rhabdomyolisis, Acidosis (H into cells, K out of the cells) and Adrenal insufficiency.

3. Benzathine Penicillin G.
- If allergic: Doxycycline, Tetracycline.
- If allergic but pregnant: do not use Doxycycline. You need to treat with penicillin after you desensitize the patient.
- If allergic with neurosyphilis: same as pregnant, desensitize!

4. Ceftriaxone unless a special case:
- Ceftriaxone and ampicillin if there is positive steroid usage, neutropenia, pregnancy, lymphoma, HIV or if pt is elderly or neonate because ampicillin covers Listeria!

Day 15   Questions 5/5

Topic: β-Hcg.

1. Where is it made?
2. When do you see the highest levels of it during pregnancy?
3. what should you think of when the levels are increased?
4. what about low then normal levels?

Topic: Retinal vein and artery blockage (both have sudden, painless, unilateral loss of vision!)

5. Describe Central Retinal Vein Occlusion:
6. Describe Retinal Artery Occlusion:

◊

7. What is the best test to Diagnose Myasthenia Gravis?
8. What are the most common side effects of Protease Inhibitors?
9. Describe Erythrasma:
10. Describe Inverse Psoriasis:
11. What cardian enzyme has the greatest sensitivity, poor specificity in detecting a cardiac insult?
12. What about best specificity?

Day 15   Answers 5/5

1. Syncytiotrophoblast.
2. Peaks at 10 weeks and the level plateaus by 20 weeks of gestation.
3. Twins or a molar pregnancy.
4. It could be a miscarraige or an ectopic pregnancy.
5. Fundoscopic exam shows disk swelling, retinal hemorrhages and cotton wool spots.
6. Fundoscopic exam shows a pale optic disc, cherry red fovea and boxcar segmentation of blood in retinal veins.
7. Electromyography.
8. Hyperlipidemia and hyperglycemia.
9. Dry, brown and scaling patches.
- Woods light positive.
10. Bright red, shiny and moist, not scaling.
- Woods light is Negative.
11. Myoglobin.
12. Troponin.

The USMLE Step Booster

1 DAY BREAK

Please take this time to review the material of the last 3 days.

SB

Day 16   Questions 1/5

1. What is CREST?
2. Describe the management of Diabetic Keto-Acidosis:
3. Describe Multifocal Atrial Tachycardia on EKG:
4. What are some causes of Multifocal Atrial Tachycardia?
5. What is the treatment of Multifocal Atrial Tachycardia?
6. A patient comes to clinic with symptoms of thyrotoxicosis, fever, atrial fibrillation, tachycardia and delirium. What is the diagnosis and treatment?

Day 16 Answers 1/5

1. Calcinosis, Raynaud's, esophageal dysmotility, Sclerodactyly, Telangiectasis.
2. Begin with IV normal saline and Insulin infusion until the blood sugar is roughly 250 mg/dL.
- Once blood sugar is roughly 250 mg/dL, NS should be replaced with D5% NS containing potassium chloride and decrease insulin diffusion rate (Insulin pushes potassium into the cells).
- Once Anion Gap stabilizes, switch IV Insulin to SC Insulin and regular food diet (give SC before disconnecting the IV line).
3. EKG will show narrow QRS, P waves or 3 or more morphologies, and variable PR segments and R-R intervals.
4. Can occure in hypoxic patients, COPD, hypomagnesemia, hypokalemia, heart disease or patients taking Aminophylline, Theophylline or isoproterenol.
5. Treatment is aimed to correct the underlying cause – Check the arterial oxygen saturation.
6. Thyroid storm.
- Treatment is PTU + propanolol + Hydrocortisone + Iodide.

Day 16 — Questions 2/5

1. A patient with a recent history of an upper respiratory infection comes with symptoms of thyrotoxicosis and painful enlarged thyroid. How would you confirm the mostly likely diagnosis?

2. A patient with malabsorption presents with fever, arthritis, and lympadenopathy. A small bowel biopsy shows PAS positive granules in macrophages. What is the diagnosis and treatment?

3. A patient presents with right lower quadrant pain and diarrhea. CT scan says that there is no appendicitis. What is the diagnose, how is it diagnosed and what is its treatment?

4. A patient with Crohn's disease presents with right flank pain rigidity that goes to the groin. What is the diagnosis?

5. What is the treatment of otitis externa?

6. Flushing, wheezing and food poisoning symptoms 5 – 15 minutes after ingesting the food? How do you treat it?

7. SLE patient is on a high dose of steroids comes to you with psychosis, delusions and hallucinations. What is the diagnosis?

Day 16   Answers 2/5

1. The most likely diagnosis is subacute thyroiditis and it is confirmed by radio iodine uptake.
- Treatment: Aspirin and propanolol.
2. Whipple's Disease due to gram positive bacilli.
- Treatment: Tetracycline or Bactrim for 1 year.
3. Yersenia enterocolitica.
- Diagnose is made by stool culture and Elisa.
- Treatment: Ceftriaxone Plus Gentamycin or Fluoroquinilones
4. Renal stones due to increased absorption of oxalate. The stones are most likely calcium-oxalate.
5. Topical antibiotics, polymyxin B plus Neomycin.
6. Scombroid poisoning. remember, it can be also a scombroid-like posoning (bluefish, dolphin or mahi-mahi).
- Treat with antihistamines only, otherwise supportive treatment.
7. Most likely Steroids psychosis but it can also be SLE induced. The difference is that in steroid psychosis auditory hallucinations are present whereas in SLE induced psychosis, visual and tacticle hallucinations are usually present.
- The next best step in management is to order markers of disease activity, if increased then psychosis most likely SLE induced.

Day 16   Questions 3/5

1. A patient is on PTU or Methimazole which, being a good physician, you know that they can develop agranulocytosis, what should you advise this patient?

    Topic: Treatment of Mycobacterium Avium.
2. When is prophylaxis indicated?
3. What drugs are best?
4. Why choose one drug over the other?

    ◊

5. If a uterus is perforated during the endometrial biopsy, what is the next step in management?
6. After performing a hysteroscopy, the patient complains of shortness of breath and on chest auscultation, crackles are noted, explain.
7. A young female comes for infertility work up. On exam, you find an indurated nodule in the cul-de-sac. What is the next step in management?
8. What is the single dose treatment of Gonorrhea?
9. What is the single dose treatment of Chlamydia?
10. A diabetic patient with migratory erythamatous rash and angular chelosis. What should you suspect? Test to order?
11. If you see "reactive lymphocytosis" in the question, what diagnoses should you keep in mind?

Day 16   Answers 3/5

1. To stop the drug if fever or soar throat develops and to visit clinic for a CBC.
2. In all HIV patients with T-cell count less then 50 cells/uL.
3. Azithromycin or Clarithromycin.
4. Azithromycin is used once a week verses clarithomycin which is used twice a day.
5. Admit the patient for at least 24 hours to make sure no complications develope.
6. The patient became volume overloaded! (they use running water to fill the uterus to be able to see).
7. Endometriosis! you need to remove it laproscopically.
8. Ceftriaxone.
9. Azithromycin.
10. Suspect glucagonoma
- Order CT of abdomen, glucagon level.
11. Mono, cat scratch disease and use of phenytoin.

Day 16　　　Questions 4/5

1. A young athletic male presents with a hematocrit of 60. What is the most likely cause of the polycythemia?
2. A polycythemic patient complains of pain and redness of hands and feet. What is the cause?
3. Describe the lab results found in Paget's disease:
4. What is the treatment of Paget's disease?
5. What is the treatment of MRSA (methicilline resistant Staph aureus)?
6. What about the treatment of VRE (Vancomycin Resistant Enterococci)?
7. What is the initial diagnostic plan for leg claudications?
8. What is the initial management if RBC casts was present in the urine?
9. What is the significance of RBC casts in the urine?
10. A paracentesis will show what type of lab values in a spontaneous bacterial peritonitis?
11. How do you treat a spontaneous bacterial peritonitis?

Day 16 — Answers 4/5

1. Androgen induced polycythemia (assume that he is on Steroids!)
2. Erythromyalgia.
   - Treatment: Phlebotomy. Aspirin can be used along with the phlebotomy.
3. Calcium is normal, Phosphate is normal, only Alkaline phosphatase is increased.
   - Hydroxyproline is elevated in the urine.
   - X-ray will show an increased bone density.
   - Perform a bone scan for diagnose
4. Only treat symptomatic patients with bisphosphonate.
5. Vancomycin.
6. Quinupristin-Dalfopristin, Linezolid.
7. Ankle/brachial arterial pressure ratio (normal is greater or equal to 1).
8. Order a renal biopsy!
9. Glomerulonephritis.
10. WBC count will be greater than 250 neutrophils.
11. Treat with cefotaxime.

Day 16　　　　Questions 5/5

1. What is the management in a pregnant patient exposed to chicken pox and who is not immunized?
2. How do you manage a pregnant patient or immunocompromised patient who develops chicken pox?
3. What diagnose should you rule out in a patient with recurring spontaneous abortions or premature delivery?

   Topic: Rheumatoid arthritis.

4. What are some contraindications for the use of methotrexate?
5. If NSAIDS are inadequate in a treatment, what can you use?
6. What is the treatment of acute exacerbations of rheumatoid arthritis?
7. When should anti-TNF be used in the treatment of rheumatoid arthritis?
8. When should gold and immunosuppressants be used in the treatment of rheumatoid arthritis?

   ◊

9. What is the treatment of Waldernstroms Macroglobinemia?
10. What is the treatment if catheter-related systemic infections are suspected?
11. A patient with pancreatitis has jaundice, increase bilirubin, ultrasound shows a dilated bile duct. Most likely diagnose?

Day 16   Answers 5/5

1. Give Varicella zoster immunoglobulin.
2. Air born isolation and supportive care only.
3. Antiphospholipid Syndrome (SLE).
4. Do not use for rheumatoid arthritis in pregnant patients, HIV patients, liver disease, renal failure or patients with supressed bone marrow (so not in anyone who's immune system is already weak).
5. Add hydroxychloroquine.
6. short course of prednisone.
7. In moderate to severe cases if methotrexate and NSAIDS fail or are contraindicated.
8. When anti-TNF fails.
9. Emergency electrophoresis.
10. Remove the catheter and give empiric therapy with vancomycin and gentamycin.
11. Gallstone causing pancreatitis.

Day 17   Questions 1/5

1. Patient comes to clinic with symptoms of pancreatitis but his amylase is normal with lipase pending. His triglycerides are greater than 1000. Most likely diagnosis?

2. A patient with history of generalized tonic clonic seizures who is well maintained with treatment asks you when can he stop taking these medications. What should you tell him?

   Topic: Treatment for each following co-morbid conditions:

3. Hypertension and chronic heart failure:

4. Hypertension and gout:

5. Hypertension and Benign Prostatic Hyperplasia:

◊

6. List examples of drugs that would decrease the excretion of theophyllin cause its toxicity:

7. What is the next step in management in a patient whose chest x-ray showed diffuse lucency with crowding of the bronchi (tram-tracking) with cross sectional ring enhancement?

8. Describe the two types of Bronchiectasis:

9. After a positive CT scan for a diffused bronchiectasis, what next?

10. What if all the tests are negative and no cause is found?

Day 17 — Answers 1/5

1. Hepatitis due to increased trigylcerides. Amylase can be normal in such cases.
2. The patient should be seizure free for at least three years.
- Tapering of drugs will be guided by EEG.
3. Ace-Inhibitors.
4. Anything, just try to avoid Diuretics.
5. An alpha-blocker (ie. Doxzasine).
6. Erythromycin and Cimetidine.
7. The next step would be to obtain a CT scan. If positive for bronchiectasis, it will show dilation of medium and small size bronchi throughout both lungs.
8. There are two types of bronchiectasis:
- Focal bronchiectasis is due to severe pulmonary infection destroying the bronchi. Surgery is the treatment of choice in such cases.
- Diffused or generalized bronchiectasis is seen in patients who cannot fight an infection (ie. CF, IgG deficiency, immobile celia, etc)
9. Perform a sweat chloride test, serum immune globulin, ciliary motion test or skin test for aspergillosis.
10. Perform HIV test.

Day 17   Questions 2/5

1. An adrenal tumor is accidently discovered on CT scan. What is the next step in management?

2. A pregnant patient presents with an increased T4, normal free T4, normal TSH. What is the most likely diagnosis?

3. A chronic dialysis patient complains of tingling and numbness of the right hand. On examination, you note atrophy of his thenar prominence on his right hand and you get a positive Tinel's Sign. What is the most likely diagnose?

4. When you see a patient with a history of diarrhea, cramps, malabsorption and iron deficiency anemia (or osteomalacia or dermatitis herpatiformis), what disease should you think of?

5. A patient with celiac sprue was started on a gluten free diet several years ago. Now, after years of being symptoms free, he complains of abdominal pain and malabsorption regardless of his diet. What is the most likely diagnosis?

6. A hepatitis B patient complains of foot drop and rash. What is the most likely diagnosis and treatment?

7. A young female patient complains of amenorrhea, high globulin levels (increase total protein but normal albumin), positive anti-smooth muscle antibodies, positive ANA. Most likely diagnose? Confirm by? Treatment?

Day 17 — Answers 2/5

1. If the tumor is greater then 4 cm, do surgery to remove it.
- If the tumor is less then 4 cm, check urine cortisol, catecholamines, rennin and aldosterone. If it shows that the tumor is active, remove it. If labs are negative, get a CT every 6 months.
2. This patient has an increased T4 because of the increased thyroglobulin caused by pregnancy (normal).
3. Most likely carpel tunnel syndrome due to dialysis related amyloidosis; deposition of B2-microglobulin.
4. Celiac Sprue, a small intestine problem. (remember: iron and Calcium are absorbed in the small intestine).
5. Intestinal lymphoma.
6. Polyarteritis Nodosa.
- Treatment: steroids
7. Autoimmune Hepatitis.
- Confirm by: Liver Biopsy.
- Treatment: Prednisone.

Day 17 Questions 3/5

1. A patient on anti-epileptic medications wants to get pregnant. Is there any precautions?
2. Young female missed 1 contraceptive pill, what is the best advice to give to this patient?
3. Same scenario as above, but instead of one pill, the patient missed 2 or more pills, what is the best advice to give to this patient now?
4. A very ill patient with pneumonia is intubated, septic and is on vasopressin presents with low T4, low TSH, low Free T4. What is the most likely diagnosis?
5. If your patient has pneumonia that does not get better with antibiotics, what should you consider? Treatment?
6. Describe the criteria for a diagnoses of Acute Respiratory Distress Syndrome (ARDS):
7. What are some examples of low molecular weight heparin?
8. A patient presents with heavy bleeding after having a tooth extracted; PT, PTT, BT are all normal. What deficiency does this patient have?
9. A patient with a history of Lupus is pregnant. If she has no prior history of fetal loss, what preventive measure should be taken?
10. If she has had a prior history of fetal loss, what preventive measure should be taken?

Day 17 — Answers 3/5

1. Give only one anti-epileptic medication and avoid valproic acid.
2. Ask her to take 2 pills the next day and continue the pill cycle.
3. Ask the patien to use other means of contraceptive until she goes back on the pill for a week. (she can take the pills with the regular doeses and schedule whenever she remembers, but condoms must be used till then).
4. Sick Euthyroid Syndrome.
- No need to treat the thyroid or do any further investigations.
5. Wegners Granulomatosis.
- Treat with cyclophosphamide and steroids.
6. ARDS:
- A pulmonary capillary wedge pressure less then 18 mmHg.
- PaO2 to FiO2 ratio of 200 mmHg or less independant of PEEP value.
- Diffuse bilateral infiltrates on chest x-ray (but unlike heart failure, ARDS patients usually have a normal clear lung exam)
7. Enoxaparin, dalteparin, nadroparin.
8. Factor 13 deficiency.
9. Put her on Aspirin throughout her pregnancy.
10. Put her on Aspirin and Heparin.

Day 17　　　Questions 4/5

1. An elderly patient presents with ecchymoses over dorsal surface of both hands; bleeding studies are normal. What is the most likely diagnosis?

2. A young female comes with complaint of fever, neck stiffness and a headache (so you suspect meningitis)/ She also presents with vaginal discharge, painful erythematous lesions on genitalia and enlarged tender inguinal lymph nodes. Lumber puncture CSF shows increased lymphocytic wbc count, normal glucose and a mild elevation in protein. What is the most likely diagnosis and treatment?

3. What is the prophylaxis for Meningitis? What is the prophylaxis for meningitis if the patient is taking an oral contraceptive pill?

4. What is the treatment of psoriatic lesions? List some medications that can exacerbate psoriatic lesions:

5. What is the treatment of Kallman's Syndrome?

6. A Patient with chronic Hepatitis presents with nausea and vomiting, jaundice, weakness, with an increased in AST and ALT (in the thousands), what is the most likely diagnose? How will you Confirm it? How would you treat it? What would be the treatment if levels are greater than 200?

Day 17   Answers 4/5

1. Benign senile purpura; increased fragility of superficial vessels due to aging.
2. Most likely genital herpes causing aseptic meningitis.
- Treatment: Acyclovir.
3. Rifampin.
- If the patient is taking on oral contraceptive pill then: Ciprofloxacin single dose
4. Treatment of psoriatic lesions: Betamethasone (if positive for athritis and or involves nails, treat with Methotrexate)
- B-blocker, NSAIDS, ACE Inhibitors, Lithium, Ant-malarials
5. Testosterone injection and if they desire to have children, treat with GnRH.
6. Acetominophin Toxicity
- Confirm by obtaining an Acetominophen level
- Treatment would be to Lavage with activated charcoal
- If Acetominophen greater then 200, Use Acytelcystine.

Day 17   Questions 5/5

Topic: CHF!

1. A patient presents with symptoms of chronic heart failure: shortness of breath, lower extremity edema, bilateral crackles. Describe the initial work up:

2. What is the initial treatment and discharge plan for the above patient?

◊

3. How would you differentiate between primary and secondary Raynaud's phenomenon?

Day 17 — Answers 5/5

1. O2 via nasal canula.
- Check pulse Oximetry (if desaturating, intubate!)
- Cardiac monitors, if abnormal, do 12-lead EKG.
- IV line, draw all routine blood including cardiac enzymes.

2. Start diuretics IV
- If in distress: Morphine IV
- If blood pressure greater then 180/110 starts IV Nitrates
- Reassess in 15-20 minutes
- Before treating with ACE-Inhibitor, check Potassium and creatine levels.
- Check intervally. If stable, send to floor
- Echo and BNP levels may be obtained.
- Discharge on oral dieuretics (ACE-I or spironolactone plus Beta-Blocker)

3. Do a nail fold capillary microscopy:
- If distorted capillary loops then it is a connective tissue disease
- If loops are normal then it is Primary Raynaud's and no further work up is needed.

Day 18   Questions 1/5

1. A young female is brought to the emergency room after being in a motor vehicle accident. She presents with stable vitals but complains of mild pain and on physical exam, you notice abdominal bruises. Describe the initial work up:

2. An HIV patient presents with fever, cough, shortness of breath, crackles in the right lower lobe (RLL), CXR shows RLL infiltrate and his CD4 count is 350. What is the most likely diagnose?

3. An HIV patient presents with fever, cough, SOB, bilateral diffused crackles and rhonchi. CXR shows bilateral diffused infiltrate and a CD4 of 150. What is the most likely diagnose? How would it be confirmed by? What is the treatment?

4. If someone accidentally got stuck by a needle from an HIV patient or had unprotected sex with an HIV patient, what treatment if any is recommended?

Day 18 Answers 1/5

1. Focused physical exam
- IV line, CXR, CBC, cross-match, BMP, pregnancy test.
- Give Morphine if she complains of pain
- If CXR negative and spleen injury is suspected order CT of abdomen.
- Send to ICU.
2. Community Acquired Pneumonia.
3. The most likely diagnose is Pneumocystis Carnii. Patients are often severely hypoxic with an A-a gradient that may be high.
- Confirm; Sputum for Methenamine silver or Right Giemsa stain.
- Treatment: IV Bactrim or Pentamidine if allergic.
- If PaO2 is greater than 70 or A-a greater than 35, add Steroids.
4. One month of 2 nucleosides and a protease inhibitor.

Day 18   Questions 2/5

Drug-induced Anemia

1. What labs would be expected in anemia caused by Alpha-Methyldopa?
2. What labs would be expected in anemia caused by Penicillin?
3. What labs would be expected in anemia caused by Quinidine?
4. What lab values will you see in Hemolytic Anemia?

◊

5. A 40 yr old female presents with headache, decreased libido, decreased testosterone, decreased LH & FSH, and a prolactin level of 800. Physical exam shows bitemporal hemianopia. What is the most likely diagnose? How would this diagnose be confirmed? What is its treatment?
6. A patient with a prolactinoma becomes pregnant. What is the next step in management?
7. What are some complications of Acromegaly?
8. A young female presents with a urinary tract infection. What is the best advice to give her?

Topic: HIV meds.

9. What HIV drug combination is unacceptable?
10. What combinations should be avoided?

Day 18   Answers 2/5

1. Direct Coombs (+) for both IgG & C3
- Indirect coombs + without adding drug.
2. Direct Coombs (IgG +, C3 +)
- Indirect only (+) when adding drug.
3. Direct is only (+) for C3
- Indirect only (+) when adding drug.
4. Increased reticulocyte count, increased LDH, Increased Indirect Bilirubin, decreased Haptoglobin.
5. Prolactinoma, confirm by MRI, Treatment: Bromocriptine or cabergolin (dopamine agonist).
6. Stop bromocriptine or cabergolin. Follow up with regular visual field exam NOT prolactin level (pregnancy increases prolactin levels).
7. Pseudogout, carpal tunnel syndrome, sleep apnea.
8. Make sure she knows the whipping technique after urinating (front to back).
9. AZT (Zidovudine) Plus d4T (Stavudine). These drugs antagonize each other.
10. DDI and DDC since both have same side effect (pheripheral neuropathy)
- AZT and Gancyclovir since combination increases risk of Bone marrow suppression
- D4T and DDI in pregnant patients since combination increases risk of lactic acidosis.

Day 18 — Questions 3/5

1. Before resecting a lung tumor, what must you do?
2. A patient presents with fever, polyuria and dysuria. On PE you note severe costovertebral angle tenderness. What is the most likely diagnose? What type of work up would you order? If the patient is pregnant, What is the most likely diagnose and describe the work up?
3. What test would you order for recurrent gonococcal bacteremia?
4. A patient on the hospital floor develops chicken pox and a nurse calls you for advice because she is seronegative for varicella. What would you tell her?
5. An infant with the history of conjunctivits presents with pneumonia. What is the most likely diagnose?
6. A premature infant presents with conjunctivitis after two weeks from birth. What is the most likely diagnose? How is it confirmed?
7. A child is brought in by his mother with the complaint that since morning her child has been experiencing visual difficulty and falling. On further history taking, you are told that he had a case of chicken pox 2 weeks ago. What is the most likely diagnose?

Day 18   Answers 3/5

1. Do Pulmonary function test: Vital Capacity must be atleast 50% of 2L.
2. Most likely Pyelonephritis. Get Urine culture and start same drugs for UTI for 10 to 14 days. If the patient has systemic symptoms (nausea, vomiting, decreased BP, fever, increased Bun/Cr) you must admit. Do urine, blood culture and renal ultrasound. Start IV antibiotics: Cipro, Levo, ceftriaxone…etc
- If the patient is pregnant, always admit and treat with Ampicillin plus Gentamicin.
3. A CH50 level which is the screening test for complement deficiency. If it comes back decreased compared to normal values, check terminal complement levels (C4, C6, C7, C8).
4. The nurse can be vaccinated but it takes 4 to 6 weeks for seroconversion (to work).
5. Chlamydial pneumonia which is based on clinical suspicion. Treatment is Erythromycin.
6. Chlamydia conjunctivitis. Confirm by PCR.
7. Most likely: post-infection cerebellar ataxia.
- Can occur after having chicken pox or measles.

Day 18 Questions 4/5

1. A Child presents with otitis media, what is the treatment?
2. A Child presents with otitis media and conjunctivitis, what is the treatment?
3. A 50 year old female comes for her routine exam and is found to be positive for VDRL. She is asymptomatic, does not recall any genital lesions or rash. What is the most likely diagnose and its treatment?
4. A patient presents with painless hematuria, UA shows WBC and RBC but no cast. IVP shows multiple areas of uretic strictures (like rosary beads) What is the most likely diagnose?
5. What is the treatment of alcohol withdrawal?
6. What is the most common cause of viral meningitis and encephalitis in children?
7. What is the most common cause of viral meningitis and encephalitis in adults?
8. What is the treatment of CML?
9. A Perimenopausal Patient presents with irregular, heavy, and dysfunctional bleeding. What is the next step in management?

Day 18　　　　　Answers 4/5

1. Amoxicillin (70% cure rate). If no improvement occurs within 24 hrs treat with erythromycin AND a sulfa drug.
2. Amoxicillin alone is not going to help, you must add Clavulinic acid. Most likely organism is H. Influenza.
3. Most likley Latent Syphyllis.
- Treatment: Benzathine Penicillin once a week for three weeks, if allergic treat with Doxycycline for one month.
4. Renal Tuberculosis; send urine for acid fast bacilli.
5. Benzodiazepine Plus thaimine, folate and multivitamins
- Add Haloperidol for hallucinations
- Add Beta-blocker for Hypertension and Tachycardia
6. Arbovirus and Enterovirus. If in a rural setting, think of Arbovirus since it is transmitted through animal vectors.
7. Herpes Simplex.
8. Imatinib Mesylate.
- Its side effects is nausea and periorbital swelling.
- If unable to tolerlate, use hydroxyurea, alpha-interferon.
- For acute blast crises use cytarabine and daunorubicin
9. Obtain a vaginal ultrasound (to assess endometrial thickness)
- Normal is less than 4mm.
- If greater than 4mm, obtain an endometrial biopsy.

Day 18　　　Questions 5/5

1. Describe the step wise management of vaginal discharge or pruritis:
2. You think your patient has appendicitis so you send the patient to surgery. To your surprize, you note that the appendix is fine! what do you do now?
3. What does amaurosis fugax indicate (what's its significance)?
4. What do you do with a patient diagnosed with amaurosis fugax?
5. Describe benign glycosuria of pregnancy:
6. In a pregnant patient suspected of having hyperthyroidism, what test will you order to screen this patient?

Day 18　　　　　Answers 5/5

1. Urine Analysis
- pH of Vaginal fluid
- Hanging drop with saline and KOH preparation
- Whiff test (optional)

2. If the patient has crohn's disease but the cecum is not involved (not inflamed): do appendectomy even in the absence of appendicits (do it anyway!).
- If the cecum is involved, do not perform the appendectomy.

3. The patient has carotid atheroslcerosis and most likely has a carotid bruit.

4. Obtain a carotid doppler evaluation to see if a carotid endarterectomy is needed.

5. Common finding because of decreased renal threshold for glucose levels.
- Other causes must be ruled out to diagnose.

6. SSRI, (ie. Fluoxetine), if resistant then treat with Alprazolam.

The USMLE Step Booster

Day 1 — Questions 1/5

1 DAY BREAK

Please take this time to review the material of the last 3 days.

SB

Day 19   Questions 1/5

1. What is the drug of choice in the emperic treatment of a patient with cystic fibrosis and severe exacerbation of pulmonary disease?

2. Your patient presents with trauma to the head and brief loss of consciousness, begins to complain of headache, vomiting and weakness of the limb contralateral to side of head injury with ipsilateral occulomotor nerve compression. What do you think the diagnosis is?

3. How do you manage such a patient?

4. A patient presents with weakness, central obesity and easy bruisability. Labs show low serum Potassium, 24 hr urine free cortisol is significantly increased and ACTH is decreased. What is the next best test?

5. What is SMR and how do you calculate it?

Day 19   Answers 1/5

1. An Aminoglycoside with an antipseudomonal penicillin (pipercillin, tobramycin).
2. The diagnosis is most likely an Acute Epidural Hematoma.
- It will show as a convex bleeding on CT scan.
- Remember: the hemorrhage originates most often from a branch of middle meningeal artery.
3. Management:
- Intubation and hyperventilation to PCO2 25-30.
- IV mannitol.
- Maintain systolic > 100 mm hg.
- Stress ulcer prophylaxis with PPI, H2 blocker or sucralfate
- Admit to ICU.
4. ACTH is low and cortisol is high: Do a CT of abdomen for adrenal mass.
- If ACTH was increased too, a CT of the brain would have been the next test (only lung or pituitary problem would cause this).
5. SMR stands for "Standard Mortality Rate"
- Observed number of deaths divided by expected numbers of deaths.
- It represents an adjusted measure of overall mortality and is adjusted for age.
- Used in occupational epidemiology.

Day 19   Questions 2/5

1. Describe the diagnostic plan in cases of renal artery stenosis:

    Topic: HIV Transmission.
2. Vaginal transmission, female to male:
3. Vaginal, male to female:
4. Needle stick transmission:
5. Anal intercourse (receptive):

    ◊

6. What test is used to screen for hemochromatosis?
7. What test confirms hemochromatosis?
8. What is the treatment for hemachromatosis?

Day 19 — Answers 2/5

1. Renal Ultrasound: would show one kidney smaller than the other.
   - You also need one of the following:
   - Captopril Renogram: blocks the effect of renin on the kidney so flow will diminish mostly on the affected kidney.
   - MRA (magnetic resonance angiography)
   - Duplex Ultrasound of the renal artery
   - Renal Artery Arteriography (Gold Standard).
2. one in 3000 exposures.
3. one in one thousand exposed.
4. one in three hundred exposed.
5. one in one hundred exposed (highest odds).
6. Transferrin saturation greater than 50% with an elevated ferritin level.
7. Gene study.
8. Phlebotmy is the treatment of choice and is begun when ferritin is greater than 200 in pre-menopausal woman or greater than 300 in menopausal and post menopausal patients
   - Deferoxamine is used if patients cannot tolerate phlebotomy.

Day 19   Questions 3/5

1. What is the effect of Amioderone and Lithium on the thyroid?
2. What is the treatment of myxedema coma?
3. How should the drug dosage be adjusted in hypothyroid patients who become pregnant?
4. What are some causes of isolated elevation in serum creatine?
5. A patient comes to you with complaint of symptoms similar to Mono however the monospot test is negative, what is the next step in management?
6. What is the most effective treatment for borderline personality?
7. What is the most specific test to diagnose multiple myeloma?
8. What is Xanthocromia?
9. When should Nimodipine be used?
10. If the ratio of WBC to RBC is greater than 1 to 500 in the CSF, what should you suspect?

Day 19 — Answers 3/5

1. Amiodarone can cause both hypothyroid and hyperthyroid.
   - Lithium can only cause hypothyroidism.
2. IV hydroxocortisone and IV levothyroxine.
3. Synthroid should be increased by 30 percent (one extra pill every two days) before even checking the TSH – increase first then check level in 4 weeks.
4. Can be caused by decreased tubular secretion of creatinine.
   - Other considerations: cimetidine, probenicid, trimethoprim
5. Obtain an EBV-specific Antibody test.
6. Dialectical behavior therapy.
7. Bone marrow biopsy and aspiration.
   - If it shows greater than 30 percent plasma cells, it is multiple myeloma.
8. A yellow discoloration of the CSF that can occur after breakdown of RBC's (tells you that the blood is from a hemorrhage, not from the procedure).
9. It is a Ca-channel blocker; used in cases of subarachnoid hemorrhage to prevent vasospasm.
10. Meningitis.

Day 19   Questions 4/5

1. List three causes that would increase Ca but decrease PTH:
2. What is the most common side effect of Allendronate, a bisphophonate?
3. What is the screening test for cushings Syndrome?
4. What is the treatment of Cushings?
5. What is the major side effect of sulfasalazine?
6. If a patient cannot tolerate methotrexate in treating Rheumatoid arthritis, what is the next best treatment?
7. When should you use Etanercept, or Infliximab in treating Rheumatoid arthritis?

Day 19   Answers 4/5

1. Malignancy (squamous cell) of lung, sarcoidosis, Vitamin D intoxication.
2. Gastritis
- Advise patient to take on empty stomach and to not lay down for 30 minutes after intake.
3. 24 hour urine of free cortisol (cushings = increased cortisol).
4. If the patient is not a surgical candidate, treat with ketoconazol.
5. Impairs folate absorption (supplement with 1 mg folic acid a day).
- Also it can suppress bone marrow; therefore, must check complete blood count.
6. Try Leflutamide, must monitos liver function test.
7. They are TNF inhibitors; add to methotrexate if disease is not well controlled (these drugs increase risk for lymphoma).

Day 19   Questions 5/5

1. What is the work up in a patient with suspected hyperthyroidism?
2. Side Effect of Foscarnet?
3. HIV patient with centrally umbilicated papules (dome shaped) lesions on the skin, what is the most likely diagnose?
4. Side effects of inhaled steroids?
5. Side effects of beta2 agonists?
6. What is the Management of a pregnant patient with a herpes infection?

Day 19   Answers 5/5

1. First check thyroid function (TSH, T4...etc)
- If Hyperthyroidism confirmed, get radio active iodine uptake:
   - Diffuse increase in iodine uptake: Grave's Disease.
   - Focal increase: Toxic Adenoma.
   - Decrease uptake can be two things:
      (1) Factitious: small nontender gland, thyroglobulin normal.
      (2) Thyroiditis: Gland is large and tender, thyroglobulin is increased.
2. Nephrotoxic, Hypocalcemia.
3. Most likely diagnoses: Molluscum contagiosum or Cryptococcus. Get cryptococcal antigen and skin biopsy.
- If Molluscum, treat with curattage or cryotherapy with liquid nitrogen.
- If Cryptococcal, treat with fluconazole.
4. Hoarse voice, sore throat, oral candidiasis.
5. Palpitations; can precipitate myocardial infarction in coronary artery diseased patients.
6. You need to treat with Acyclovir starting at 36 weeks until delivery
- If the patient has vulvular pain or vesicles while in labor, proceed to Caesarean section.

Day 20 Questions 1/5

1. What is the typical Presentation of tubuerculosis in children?

    Stabbed in the neck!

2. if on physical exam you noticed or suspected an expanding hematoma, what should you do?
3. What if you felt or heard crepitations in the neck?
4. What if the stabbed patient is a child younger than 13 years old?

    ◊

5. What is the best test to screen for acromegaly?
6. Patient with candidal vaginitis has frequent relapses, what is the next step in management?

Day 20   Answers 1/5

1. In children, typically No cavitary lesion is seen on x-ray, is non-infective and often affects the lower lobe similar to pneumonia.

2. Expanding hematoma: protect the airway by orotracheal intubation.

3. If crepitations are present, do not blindly intubate! first perform a bronchoscopy then the orotracheal intubation can be performed.

4. In young children, you should never perform any sort of intubation. Oxygen should be inserted into the trachea via a large bore needle (needle crichothyroidotomy) then a tracheostomy can be performed.

5. Increased insulin-like growth factor 1.

- To confirm: oral glucose suppression test, then MRI or CT to locate pituatary tumor.
- Treatment: Surgery (transphenoidal)

6. Order blood glucose and HIV test.

Day 20   Questions 2/5

1. A patient is in labor, and you notice herpetic lesions on the face, what is the next step in management?
2. A patient has genital herpes and is in labor with ruptured membranes, what is the next step in management?
3. Patient being worked up for dementia and is found to be positive for VDRL. What is the next step?
4. What is the treatment of Lichen sclerosis?
5. An Asian woman presents with a low MCV and normal iron level. What is the next best test to order?
6. What are some causes of Pulsus paradoxus?
7. What types of shock gives you low SVR (systemic vascular resistance)?
8. What effects does sickle cell trait have on a pregnant patient?
9. A pregnant patient had a case of UTI, you treated with antibiotics, what is the next step in management?
10. If follow up is again positive, what is the next step in management?
11. A pregnant patient comes with pyelonephritis, next step in treatment?

Day 20   Answers 2/5

1. No need for Caesarean section (only required if lesions are present on the genitals).
2. Caesarean section.
3. Send cerebral spinal fluid for VDRL to rule out neurospyhillis.
4. High potency corticosteroids.
5. Rule out Thallesemia (always think of this when an asian woman presents with low MCV).
6. Cardiac Tamponade, Asthma, COPD.
7. Only Septic shock and Anaphylactic shock!
8. Those patients have an increased risk of urinary tract infection, otherwise the pregnancy should be uncomplicated.
9. Follow up urine culture.
10. Treat with another course of antibiotics (depending on sensitivity).
11. Admit the patient.
- This patient should receive suppressive therapy (antibiotics) for the entire pregnancy (the same goes for patients with repeated urinary tract infections).

Day 20   Questions 3/5

Topic: More about HIV treatment.

1. What is the next treatment/advice for an HIV patient with CD4 count of 30 on triple treatment and bactrim?
2. How do you treat an HIV patient with thrombocytopenia?
3. How do you treat an HIV patient with Anemia?
4. How do you treat an HIV patient with severe neutropenia?
5. An HIV patient is exposed to the chicken pox who has no prior exposure with negative antibodies present, what is the next best management?
6. A patient is on protease inhibitors and develops an increase in his LDL, what is the next step in management?
7. If the same patient has triglycerides greater then 500 mg/dl, what is the next step in management?

◊

8. During labor, the patient suddenly has shortness of breath and a decrease in blood pressure, what is the most likely diagnosis?
9. What antipsychotic drug is safe for pregnancy?
10. What about for depression?
11. Can lactating mothers take TCA (Tricyclics Antidepressants)?

Day 20   Answers 3/5

1. MAI prophylaxis: Azithromycin or clarythromycin.
2. Just the anti-retroviral therapy, no additional medications necessary.
3. Treat with erythropoietin (anemia of chronic disease).
4. Treat with granulocyte colony stimulating factor.
5. Give Varicella Zoster Immunoglobulin.
6. Pravastatin / Atorvastatin.
7. Add Gemfibrozil.
8. Amniotic fluid embolism, if she survives, she is at risk for disseminated intravascular coagulation.
9. Haloperidol.
10. SSRI: Fluoxetine.
11. Yes, it's safe.

Day 20   Questions 4/5

1. What is Atrophic vaginitis and how do you treat it?
1. What is the difference between thrombotic thrombocytopenia purpura (TTP) and hemolytic uremic syndrome (HUS)?
2. What is the treatment for TTP / HUS?
3. Chest x-ray displays cavitation in a pneumonia patient. What are the most likely organisms?
4. Patient with fever, weight loss, cough that has foul smelling sputum. On observation, you notice poor dental hygiene. Chest x-ray shows infiltrates present in dependent lung zone. What is the most likely cause?
5. What is the typical presentation of Mycoplasma Pneumonia?

Day 20　　　　　Answers 4/5

1. Occurs in elderly postmenapausal women. presents as watery yellowish vaginal discharge and dyspareunia. vagina is thin and pale.
- It is basically dryness and inflammation of the vagina due to the lack of estrogen.
- Wet mount would show numerous WBC, no bacteria, KOH would be negative.
- Treat with topical Estrogen.
2. HUS patients do not have neurologic symptoms and they have definite renal failure, whereas TTP may or may not have renal failure.
3. Emergency plasmaphoresis.
- Only supportive treatment for children with HUS.
4. Staph aureus (Intravenous drug abusers), Klebsiella (Alcoholics or Diabetes Mellitus) or Pseudomonas.
5. Aspiration pneumonia, seconday to anaerobes.
6. Treat with Clindamycin.
7. Similar question can say that the patient is an elderly home residnet with frequent choking spells.
8. Patients are usually young/college students.
- Must have cold antibody associated hemolytic anemia.
- Treatment is Erythromycin or Azithromycin.

Day 20 Questions 5/5

1. Patient with shortness of breath, increase jugular venous pulse, hypotensive, has paradoxical pulse of 22 mmHg. EKG shows amplitude changes with every beat (electrical alternans) and low voltage. What is the most likely diagnose?

   Topic: A little more about HIV.

2. How do you check the HIV status of a newborn from an HIV positive mother?
3. If HIV patient is on bactrim for PCP prophylaxis and now his CD4 count is less then 100, what drug do you add for prophylaxis against Mycobacterium Avium and Toxoplasmosis?
4. What if the same patient was on dapsone instead of bactrim?

5. What is the indication for penumoncoccal vaccine?
6. What is the indication for the Influenza vaccine?
7. What is sialolithiasis?
8. What is Sialadenitis?

Day 20   Answers 5/5

1. Cardiac Tamponade.
- Treat with performing pericardiocentesis.
- To confirm, perform echocardiogram.

2. Order "HIV DNA PCR"
- Any other test would show the newborn to be positive even if he is negative.

3. Add Clarithromycin or Azithromycin (for M. Avium) because Bactrim covers Toxoplasmosis.

4. In this situation, you must add clarithromycin or azithromycin for M. Avium and Pyrimethamine for Toxoplasmosis. Dapsone alone does not cover Toxoplasmosis.

5. Patients greater then 65 years old or patients at high risk (ie. Splenectomy, HIV, Immunocompromised patients on chemotherapy or post-transplants).

6. Give annually for all patients greater then 50 years old and high risk patients.

7. Calculus formation in Wharton's duct (submandibular gland) or the Stenton's duct (Parotid gland).

8. It is the inflammation of the salivary glands, usually the parotid and submandibular glands associated with acute swelling and pain with meals.
- If viral think of: EBV, Coxsackie, Parainfluenza
- If Bacterial think of: Staphlococcos Aureus

The USMLE Step Booster

2 DAY BREAK

Please take this time to fill in the next three pages with your own notes of what you think is necessary to memorize for the exam. Then go back and review day 19 and day 20 along with your notes.

The USMLE Step Booster

Notes

Notes

The USMLE Step Booster

Notes

The USMLE Step Booster

Congratulations!

You did it!
Best of Luck on the examination!

www.ingramcontent.com/pod-product-compliance
Ingram Content Group UK Ltd.
Pitfield, Milton Keynes, MK11 3LW, UK
UKHW021319180426
11947UKWH00015B/1322